T0288847

Adding Value to Air Force Management Through Building Partnerships Assessment

Jefferson P. Marquis, Joe Hogler, Jennifer D. P. Moroney,
Michael J. Neumann, Christopher Paul, John E. Peters,
Gregory F. Treverton, Anny Wong

Prepared for the United States Air Force

RAND PROJECT AIR FORCE

The research described in this report was sponsored by the United States Air Force under Contract FA7014-06-C-0001. Further information may be obtained from the Strategic Planning Division, Directorate of Plans, Hq USAF.

Library of Congress Cataloging-in-Publication Data

Adding value to Air Force management through building partnerships assessment / Jefferson P. Marquis ... [et al.].
    p. cm.
  Includes bibliographical references.
  ISBN 978-0-8330-5089-2 (pbk. : alk. paper)
  1. United States. Air Force—Operational readiness. 2. United States. Air Force—Foreign service. 3. Combined operations (Military science) 4. Military art and science—International cooperation. 5. United States—Military relations—Foreign countries. 6. Military assistance, American. I. Marquis, Jefferson P.

  UG633.A438 2010
  358.4'160973—dc22

                                                                      2010048769

The RAND Corporation is a nonprofit institution that helps improve policy and decisionmaking through research and analysis. RAND's publications do not necessarily reflect the opinions of its research clients and sponsors.

RAND® is a registered trademark.

Published 2010 by the RAND Corporation
1776 Main Street, P.O. Box 2138, Santa Monica, CA 90407-2138
1200 South Hayes Street, Arlington, VA 22202-5050
4570 Fifth Avenue, Suite 600, Pittsburgh, PA 15213-2665
RAND URL: http://www.rand.org/
To order RAND documents or to obtain additional information, contact
Distribution Services: Telephone: (310) 451-7002;
Fax: (310) 451-6915; Email: order@rand.org

Preface

The U.S. Air Force (USAF), along with other elements of the U.S. Department of Defense (DoD), has worked for many years with allies and friendly nations to build strong and enduring partnerships, reinforce other nations' capacity both to defend themselves and to work in coalitions, and ensure U.S. access to foreign territories for operational purposes. However, it is often challenging to specify just how these activities have contributed to U.S. security policy objectives, and how much or in what ways. This technical report builds on previous RAND Corporation research by helping the Air Force to implement an approach to assessing Air Force security cooperation programs that accounts for the diversity of perspectives, responsibilities, and capabilities within the Air Force's security cooperation community. We have attempted to achieve this objective by (1) obtaining the views of a broad spectrum of Air Force officials regarding assessment in general and RAND's proposed assessment framework in particular, (2) gaining insights into existing Air Force capacity to conduct security cooperation assessments, and (3) strengthening the case for a comprehensive assessment approach by (a) addressing essential "why, what, how, and who" questions and (b) applying the answers to specific Air Force programs.

This RAND Project AIR FORCE (PAF) report documents research performed for a fiscal year (FY) 2009 study titled "Implementing a Framework to Assess Programs That Build Air Force Partnerships." The work was sponsored by the Office of the Deputy Chief of Staff for Operations, Plans and Requirements, Headquarters USAF (AF/A3/5) and was conducted within the Strategy and Doctrine Program of RAND Project AIR FORCE. It is the latest in a series of PAF studies supporting the Air Force's efforts to work with partner air forces across the spectrum of operations.

Other PAF documents that address security cooperation and security cooperation issues include the following:

- *Developing an Assessment Framework for U.S. Air Force Building Partnerships Programs,* Jennifer D. P. Moroney, Joe Hogler, Jefferson P. Marquis, Christopher Paul, John E. Peters, and Beth Grill (MG-868-AF).
- *International Cooperation with Partner Air Forces,* Jennifer D. P. Moroney, Kim Cragin, Eric Gons, Beth Grill, John E. Peters, and Rachel M. Swanger (MG-790-AF).
- *Air Power in the New Counterinsurgency Era: The Strategic Importance of USAF Advisory and Assistance Missions,* Alan J. Vick, Adam Grissom, William Rosenau, Beth Grill, and Karl P. Mueller (MG-509-AF).

RAND Project AIR FORCE

RAND Project AIR FORCE (PAF), a division of the RAND Corporation, is the U.S. Air Force's federally funded research and development center for studies and analyses. PAF provides the Air Force with independent analyses of policy alternatives affecting the development, employment, combat readiness, and support of current and future aerospace forces. Research is conducted in four programs: Force Modernization and Employment; Manpower, Personnel, and Training; Resource Management; and Strategy and Doctrine.

Additional information about PAF is available on our website:
http://www.rand.org/paf/

Contents

CHAPTER FIVE

Figure and Tables

Figure

Tables

Summary

The USAF, along with other DoD elements, has worked for many years with allies and friendly nations to build strong and enduring partnerships, reinforce other nations' capacity both to defend themselves and to work in coalitions, and ensure U.S. access to foreign territories for operational purposes. The activities conducted by the Air Force range from the very visible—training, equipping, and exercising with others—to those that are less obvious, such as holding bilateral talks, workshops, and conferences and providing education.

Since 2006, DoD has placed a higher priority on these security cooperation activities, which collectively are viewed as being central to U.S. efforts to shape international relations in ways that are favorable to U.S. interests and equities.[1] As the demand for security forces continues to outpace the supply, the United States needs partners to improve their own capabilities and be better prepared to confront both internal and external security challenges. As a result, U.S. efforts to build partnerships with foreign countries have evolved from the "nice to do" category to the "necessary" one.

However, it is often challenging to specify how much and what ways these activities have contributed to U.S. security policy objectives—whether at the national, department, combatant command (COCOM), or service levels. Given this challenge, the Air Force is seeking ways to assess the effectiveness of its security cooperation efforts in order to make the best use of its resources in initiatives that are intended to enhance partner capabilities and achieve U.S. security objectives.

In pursuit of this goal, PAF has been working with the Air Force to develop an assessment approach designed to ensure that Air Force security cooperation programs and activities are closely aligned with operational and strategic objectives, adequately authorized and resourced, carefully sequenced and packaged, and efficiently and effectively executed. To be successful, this effort must overcome skepticism among some in the Air Force security cooperation community regarding the need for added and more rigorous assessments, the capacity of the Air Force to perform additional assessments given existing resource constraints, and even the goal of pursuing assessments that go beyond the current short-term, process-oriented assessments, despite the often long-term, unexpected, and indirect effects of security cooperation activities.

For this study, RAND was asked by the Office of the Deputy Chief of Staff of the Air Force for Operations, Plans and Requirements, Headquarters USAF (AF/A3/5), to help the Air Force implement the comprehensive assessment framework developed by PAF for the Office of

---

[1] Security cooperation comprises all DoD interactions with foreign defense and security establishments, including all DoD-administered security assistance that builds defense and security relationships, develops allied and friendly military capabilities for self-defense and multinational operations, and provides U.S. forces with peacetime and contingency access to host nations. See Department of Defense Directive 5132.03, *DoD Policy and Responsibilities Relating to Security Cooperation*, October 24, 2008.

the Deputy Under Secretary of the Air Force for International Affairs (SAF/IA) in FY 2008, particularly but not exclusively with respect to security cooperation programs directly managed by the Air Force.[2]

In fashioning this comprehensive framework for implementation by the Air Force, the RAND team

- tested and refined the elements of the assessment framework through structured Assessment Day focus group discussions with a variety of Air Force officials and senior leaders involved in security cooperation with foreign militaries (see Chapter Two)
- conducted a detailed survey of stakeholders in the Operator Engagement Talks (OET)[3] and the Military Personnel Exchange Program (MPEP) to gauge the Air Force's ability and capacity to conduct security cooperation assessments (see Chapter Three)
- proposed a practical approach to implementing a security cooperation assessment framework based on the Air Force's legal authorities, policy guidance, organizational responsibilities, and assessment capabilities (see Chapter Four).

RAND's 2009 study will enable the Air Force to better meet its security cooperation assessment challenges because this approach is rooted in three key assumptions:

1. Assessment benefits the things assessed.
2. Air Force efforts to build foreign partnerships can benefit if there is a common understanding of security cooperation assessment.
3. Security cooperation assessments can support DoD and Air Force decisionmaking if they are available, current, accurate, and configured appropriately.

While the first assumption is affirmed in Chapter One of the report through examples from non–security cooperation USAF endeavors, Chapters Two, Three, and Four reaffirm that assessments can benefit security cooperation efforts and support Air Force decisionmaking. The following sections explore these findings and recommendations.

Findings and Recommendations

Our second assumption states that focused, integrated security cooperation assessments will benefit the Air Force only if there is a common understanding among Air Force stakeholders of the rationale behind security cooperation assessment. Unfortunately, the results of our "Assessment Day" focus group discussions, which included Air Force security cooperation planners and executors from a variety of Headquarters U.S. Air Force (HAF), major commands (MAJ-COMs), and regional component organizations (see Chapter Two, pp. 13–20), confirmed that no consensus exists within the Air Force on four fundamental issues pertaining to security cooperation assessments: Why assess? What to assess? How to assess? And who should assess?

---

[2] That assessment framework is described in Jennifer D. P. Moroney, Joe Hogler, Jefferson P. Marquis, Christopher Paul, John E. Peters, and Beth Grill, *Developing an Assessment Framework for U.S. Air Force Building Partnerships Programs*, Santa Monica, Calif.: RAND Corporation, MG-868-AF, 2010a.

[3] This initiative was previously called the Operator-to-Operator Program or Operator-to-Operator Staff Talks.

Consequently, we recommend that SAF/IA, as the leader within the Air Force for building foreign partnerships, pilot the development of a capstone instruction to guide assessment efforts across the full range of security cooperation functions, organizations, programs, and countries in which Air Force personnel operate, either as managers, implementers, or observers. This instruction could build on the security cooperation assessment foundation outlined and illustrated in Chapter Four, which answers the four assessment questions as follows:

- **Why Assess?** The Air Force should conduct security cooperation assessments to improve partnership design and prioritization,[4] force employment, force development, and overall force management decisions (pp. 34–36).
- **What to Assess?** The Air Force should assess security cooperation events, activities, and programs, as well as their combination and interaction in the context of particular countries and multinational initiatives (p. 37).
- **How to Assess?** The Air Force should choose appropriate assessment types, objectives and effects, quantitative and qualitative measures, and relevant data (pp. 37–39).
- **Who Should Assess?** The Air Force should assign department, MAJCOM, numbered air force (NAF), wing and below, and joint elements[5] to appropriate assessment roles and clearly articulate supporting-supported relationships (pp. 39–40).

Security cooperation assessments can provide a range of specific benefits depending on U.S. and partner interests, security cooperation objectives, and force development requirements. Assessments can also inform decisions regarding the establishment, continuation, expansion, or transference of programs and activities designed to build partner capacity, foster partner relationships, or guarantee U.S. access to partner countries. Additionally, assessments can provide information relevant to generating Air Force units to perform security cooperation activities, such as training additional airmen to train, advise, and assist or expanding U.S. Air Force schools to accommodate more foreign students.

However, according to our third assumption, assessments cannot provide these benefits unless they are available, current, accurate, and appropriately configured.

Although the results of our "Assessment Day" focus group discussions and security cooperation program survey are not definitive, they point to significant problems with the Air Force's current security cooperation assessment process. Below is a brief list of our assessment process findings from Chapters Two and Three, along with a recommended course of action for the Air Force.

Assessment data are often unavailable. The Air Force should make security cooperation data collection and assessment a duty responsibility for Air Force personnel working in the field. It also should request data from joint organizations that could be used for various assessment purposes, such as the need for Air Force security cooperation resources and the efficiency and impact of Air Force security cooperation activities (pp. 17, 29).

---

[4]   Partnerships design involves selecting the appropriate combination of security cooperation events, activities, programs and resources to meet the end states articulated in U.S. Air Force, *Air Force Global Partnership Strategy: Building Partnerships for the 21st Century*, 2008, pp. ii–iii. The prioritization of partnerships involves selecting and ranking foreign partners for different missions and objectives of importance to the U.S. Air Force and the U.S. government more generally.

[5]   For the purpose of this report, "joint" is defined as Air Force personnel serving in assignments outside the Air Force, in particular, on regional COCOM staffs and in U.S. embassy country teams (e.g., air attachés and security assistance officials).

Security cooperation guidance is not consistently available or well enough understood. The Air Force should incorporate assessment guidance in security cooperation planning documents, including specific program or country objectives, measures, and data requirements. In addition, security cooperation officials should be held accountable for meeting assessment requirements (pp. 17, 30).

Assessment skills could be improved. The Air Force should consider offering a security cooperation–related online assessment course that would be available to all Air Force personnel should they wish to take it (pp. 17–18, 31).

Assessment resources may be inadequate. The Air Force should conduct an in-depth resource analysis after determining appropriate security cooperation assessment roles and supporting-supported relationships (p. 18).

Security cooperation officials lack data needed for effective program advocacy. The Air Force should ensure that data necessary for managing and advocating for security cooperation program resources are analyzed by HAF officials and used to inform management and budgetary recommendations and decisions (pp. 28–29).

Officials are often unable to compare programs. The Air Force should educate airmen and civilians working on security cooperation issues on the broader universe of Air Force Title 10 security cooperation programs by creating a handbook of Air Force security cooperation programs and activities. Alternatively, the Air Force should incorporate this information into the Air Force Campaign Support Plan as an appendix (pp. 28–29).

Implementing a New Approach to Security Cooperation Assessment

Creating a more integrated approach to security cooperation assessment should not be a burden for the Air Force. Much of the information necessary for security cooperation assessment is available, or could be available if the Air Force only ordered its collection and managed its dissemination, analysis, and integration into Air Force and DoD decisionmaking processes. The authorities seem to exist, and additional Air Force instructions could fill in certain gaps. The Air Force has forged cooperative relationships across disparate communities in the past when prudence dictated; there is no reason to believe similar relationships could not be developed in support of security cooperation assessment.

However, there are certain hurdles impeding the establishment of a new approach to security cooperation assessment. These include the doctrinal and organizational (i.e., why, what, how, who) issues, as well as the procedural, educational, and resource issues detailed above.

Those responsible for security cooperation in the Air Force need to more fully understand the dimensions of the assessment capacity problem and communicate capacity requirements to senior Air Force and DoD leaders. Toward this end, we suggest that stakeholders in Air Force–managed programs take RAND's security cooperation assessment survey and provide the results to the Office of the Deputy Chief of Staff for Operations, Plans and Requirements, Directorate of Operational Planning, Regional Affairs Division, Headquarters U.S. Air Force (AF/A5XX); SAF/IA; and Air Education and Training Command (AETC) for security cooperation planning, programming, and force development purposes.

In addition, Air Force security cooperation officials need to decide on a strategy for implementing a new assessment approach. We suggest the following four-part strategy:

1. Achieve a consensus among security cooperation stakeholders on the elements of the overall security cooperation assessment approach.
2. Codify this vision in a capstone Air Force instruction.
3. Systematically implement the new assessment approach, focusing first on Air Force–managed programs and activities that are well established, clearly defined, and adequately resourced.
4. Collaborate with other DoD and Department of State stakeholders regarding assessment policy affecting security cooperation programs and activities in which the Air Force participates but does not manage.

The question is no longer whether the Air Force should assess its efforts to build partnerships. Rather, it is how to do so in an integrated way that takes into account the evolving nature of security cooperation policy, as well as Air Force constraints with respect to assessment capacity and authority.

Acknowledgments

The authors wish to thank a number of individuals and offices for their support of the research reported in this report. First, we owe our study sponsor, Col Nancy Boriack, from Headquarters, U.S. Air Force, Office of Regional Plans and Issues (USAF/A5XX), a debt of gratitude. We also wish to thank Col Kimerlee Conner, Director of Strategy and Long Range Planning, Deputy Undersecretary of the Air Force, International Affairs (SAF/IAG). Both provided excellent feedback and research assistance during the course of the yearlong study. We are grateful for the insightful discussions we had with officials from the following agencies: SAF/IA regional and functional desk officers; Headquarters Air Education and Training Command/International Affairs; Air Force Security Assistance Training Squadron; Air Force Security Assistance Center; Air Forces Central; U.S. Air Forces in Europe; Pacific Air Forces; Air Forces Southern; Inter-American Air Forces Academy; and the air attaché community.

Abbreviations

AETC	Air Education and Training Command
AF/A1	Deputy Chief of Staff for Manpower and Personnel, Headquarters U.S. Air Force
AF/A3/5	Deputy Chief of Staff for Operations, Plans and Requirements, Headquarters U.S. Air Force
AF/A5XX	Office of the Deputy Chief of Staff for Operations, Plans and Requirements, Directorate of Operational Planning, Regional Affairs Division, Headquarters U.S. Air Force
AF/A8	Deputy Chief of Staff for Strategic Plans and Programs, Headquarters U.S. Air Force
AF/A9	Director, Air Force Studies and Analyses, Assessments and Lessons Learned, Headquarters U.S. Air Force
AFI	Air Force instruction
AFPD	Air Force policy directive
COCOM	combatant command
DoD	U.S. Department of Defense
DOTMLPF	doctrine, organization, training, materiel, leadership and education, personnel, and facilities
FY	fiscal year
HAF	Headquarters U.S. Air Force
MAJCOM	major command
MPEP	Military Personnel Exchange Program
NAF	numbered air force
OET	Operator Engagement Talks
OSD	Office of the Secretary of Defense
PAF	Project AIR FORCE
PME	professional military education
POM	program objective memorandum

PPBE	Planning, Programming, Budgeting, and Execution
RPMO	regional project management office
SAF/IA	Deputy Under Secretary of the Air Force for International Affairs
SAF/IAG	Deputy Under Secretary of the Air Force for International Affairs, Directorate of Strategy and Long Range Planning
SAF/IAP	Deputy Under Secretary of the Air Force for International Affairs, Policy Directorate
USAF	U.S. Air Force

CHAPTER ONE

Introduction

As one Air Force civil engineer explained in a discussion about the value of assessing Air Force security cooperation efforts, "We're out there building partnerships all the time, but it's not on the radarscope." Many activities take place in remote regions without support from a dedicated source of funding. According to the engineer, "We may be supporting partners at their air base, and in our free time drill a well outside the wire and get good relationships and intelligence."[1]

The engineer's comment is revealing. It underscores that the U.S. Air Force (USAF), along with other elements of the U.S. Department of Defense (DoD), has worked for many years with allies and friendly nations to build strong and enduring partnerships, reinforce other nations' capacity both to defend themselves and to work in coalitions, and ensure U.S. access to foreign territories for operational purposes. The activities conducted by the Air Force range from the very visible—training, equipping, and exercising with others—to those that are less obvious, such as holding bilateral talks, workshops, and conferences and providing education.

Yet, it is often challenging to specify how much and in what ways these activities have contributed to U.S. policy objectives—whether at the national security, department, combatant command (COCOM), or service levels. Often, as the engineer's comment indicates, the activities are not called security cooperation at all but, rather, are simply part of doing business with allies and friends. Even what the Air Force formally calls security cooperation is quite varied in focus and time, ranging from tightly focused exercises with specific goals, to education programs intended to foster long-term relationships with military officers from other nations, among many other activities and events.

The Growing Importance of Security Cooperation

Whether called security cooperation or not, these activities have evolved into a higher-priority set of tasks and chores that, collectively, are viewed as more central to U.S. efforts to shape international relations in ways favorable to U.S. interests and equities. As then–Chief of Naval Operations Admiral Michael G. Mullen acknowledged in the fall of 2005,

> [T]he United States Navy cannot, by itself, preserve the freedom and security of the entire maritime domain. It must count on assistance from like-minded nations interested in using

---

[1] The comment came during a breakout session during RAND's "Assessment Day," held on May 15, 2009, at RAND's office in Arlington, Virginia.

the sea for lawful purposes and precluding its use for others that threaten national, regional, or global security.[2]

This observation applies to the air domain as well. As U.S. interests around the world increase, while U.S. capabilities remain stable or decline, the United States needs partners to augment their own capabilities and capacity. Security cooperation has evolved from being a complementary aspect of U.S. defense strategy to a central element of that strategy.

Security cooperation planning and implementation are accomplished under various legal authorities and by different organizations. To simplify, under the Title 10 authorities of DoD, the geographic COCOMs are primarily responsible for security cooperation activities that support the employment of forces overseas. Under the State Department's Title 22 authorities, U.S. ambassadors are primarily responsible for security assistance activities in the countries to which they are accredited. In general, the Air Force and its sister services are responsible for providing to combatant commanders and ambassadors capabilities with which they can improve relationships with, build the capacity of, and increase access to foreign partner nations. Consequently, the force development dimension of security cooperation—which has recently been defined as the Building Partnerships Joint Capability Area—is especially important to the leadership of the services, a fact that the Air Force has recognized by designating "building partnerships" as a core function deserving of its own "master plan."[3] With respect to force development, the key issue is how the Air Force should generate capabilities through its doctrine, organization, training, materiel, leadership and education, personnel, and facilities (DOTMLPF)[4] functions so that it can fulfill its security cooperation as well as non–security cooperation roles and missions. In particular, are there enough airmen with appropriate language and cultural skills to interact effectively with foreign partners?

The Value of Program Assessments Is Becoming Better Understood

DoD's increased emphasis on building partnerships requires integrating security cooperation into corporate decisionmaking, which in turn requires a systematic evaluation of security cooperation needs, capabilities, processes, outcomes, and cost-effectiveness. Thus, it is little surprise that the recent emphasis on assessing the effectiveness of security cooperation activities has emerged from the suite of new DoD and Air Force planning guidance and instructions, such as the Air Force Global Partnership Strategy.[5] For the Air Force, the question is not whether to assess efforts with respect to security cooperation. The value of assessments to managing Air Force resources, enhancing DoD's capabilities, and furthering U.S. interests seems clear. Indeed, as the Air Force has been engaged in security cooperation for many years, it also has tried to assess the results of those activities in various ways. However, security coop-

---

[2] Adm Michael G. Mullen, U.S. Navy, remarks delivered at the Seventeenth International Seapower Symposium, Newport, Rhode Island, September 21, 2005.

[3] We use *force development* here in the DoD sense of the term rather than in the Air Force sense—that is, as the range of functions necessary for generating defense capabilities as opposed to simply leader development.

[4] The DOTMLPF construct is important for DoD as a whole, not just the Air Force. It is particularly useful in helping to identify service programs and resources that are needed to fulfill security cooperation goals and objectives over time.

[5] U.S. Air Force, 2008.

eration assessment has generally been narrowly focused or ad hoc, process-oriented as opposed to outcome-oriented, and inadequately organized and integrated to support Air Force and DoD decisions. Furthermore, there does not appear to be a consensus within the Air Force on the purpose, focus, conduct, or organization of security cooperation assessments. Nor is there a detailed understanding of the capacity of the Air Force—in terms of funding, personnel, resources, guidance, and expertise—to pursue a systematic approach to assessing the programs it manages and/or executes.

Study Objectives and Analytical Approach

In fiscal year 2008, RAND Project AIR FORCE (PAF) developed a conceptual framework for assessing the Air Force's security cooperation efforts at the behest of the Deputy Under Secretary of the Air Force for International Affairs (SAF/IA).[6] As a follow-on to this work, the Office of the Deputy Chief of Staff for Operations, Plans and Requirements, Headquarters U.S. Air Force (AF/A3/5), asked PAF in 2009 to help the Air Force implement a comprehensive approach to assessment, particularly but not exclusively with respect to security cooperation programs directly managed by the Air Force.

In pursuit of this goal, the RAND study team worked with AF/A3/5 and SAF/IA to better understand and overcome, if possible, certain obstacles to the implementation of RAND's proposed framework, particularly the variety of perspectives within the Air Force regarding security cooperation assessments and the lack of knowledge regarding the Air Force's ability to conduct a full spectrum of such assessments. Given the diffuse nature of authority and responsibility in the security cooperation arena, as well as the low probability of significant additional resources being applied to assessment, the PAF team believed that it made sense for the Air Force to tackle these obstacles before attempting to impose a comprehensive assessment scheme on a large part of its security cooperation community.

Accordingly, the RAND team undertook the following tasks. The team

- conducted structured discussions with a range of Air Force security cooperation officials and senior leaders regarding assessment in general and RAND's 2008 assessment framework in particular
- conducted a detailed survey of stakeholders in the Operator Engagement Talks (OET)[7] and the Military Personnel Exchange Program (MPEP) to gauge the Air Force's ability and capacity to conduct security cooperation assessments
- attempted to strengthen the case for a comprehensive approach to assessing Air Force efforts to build partnerships by modifying and applying RAND's initial assessment framework.

The RAND study team engaged in three separate efforts to collect relevant insights and data about the potential for enhancing Air Force security cooperation assessments. First, the

---

[6] That assessment framework is summarized in Chapter Two. For a more complete description, see Jennifer D. P. Moroney, Joe Hogler, Jefferson P. Marquis, Christopher Paul, John E. Peters, and Beth Grill, *Developing an Assessment Framework for U.S. Air Force Building Partnerships Programs,* Santa Monica, Calif.: RAND Corporation, MG-868-AF, 2010a.

[7] OET was previously called the Operator-to-Operator Program or Operator-to-Operator Staff Talks.

team held focus groups during the Air Force's Combat Support Plan Conference, held at RAND in May 2009, with a broad representation of nearly 40 Air Force security cooperation stakeholders. The results of those focus groups were analyzed both quantitatively and quantitatively (as detailed in Chapter Two). The team met with a former Secretary of the Air Force to get a senior leader's perspective on security cooperation and ways to assess Air Force–managed programs for decisionmaking purposes. RAND also conducted detailed surveys of OET and MPEP stakeholders to better understand the ability and capacity of Air Force service and joint personnel at various levels to undertake a full range of security cooperation assessments and assessment responsibilities (see Chapter Three). This endeavor generated the majority of the data presented in this report.

Following an analysis of these data collection efforts and further reflection on ongoing Air Force efforts to establish building partnerships as a core service function, the RAND team modified its initial assessment framework to focus on four elemental issues: Why assess? What to assess? How to assess? And who should assess? (see Chapter Four). In presenting the refined framework in this volume, we explain the connections between different kinds of assessment and major Air Force decisionmaking processes and provide examples of how RAND's revised assessment approach might be applied to specific Air Force programs (i.e., OET and MPEP).

Security Cooperation Assessment Assumptions

A fair amount of controversy attends the issue of security cooperation assessment. Skeptics argue that enough assessment goes on today—for example, in the form of the events that are reported and aggregated in the COCOMs' Theater Security Cooperation Management Information Systems and the program justifications that are compiled by each of the services as part of DoD's Planning, Programming, Budgeting, and Execution (PPBE) process. Others doubt that additional assessments would reveal anything that is not being currently communicated to decisionmakers via country desk officers in service headquarters and COCOM staffs as well as military representatives on embassy country teams. Still others note the elusive nature of security cooperation, especially in the way security cooperation activities can produce unexpected and indirect payoffs for the United States over the long term, and that these benefits are not amenable to standard evaluation techniques, which rely on short-term, quantitative measures of success.

In answer to the skeptics, we would argue that much of the assessment information that is currently being collected regarding security cooperation activities is of limited quality (e.g., focused on program inputs versus outputs and outcomes); is not being reported and analyzed in a clear and systematic manner for decisionmaking purposes; and fails to adequately address issues related to long-term effectiveness primarily because of inadequacies in collection, reporting, analysis, and integration, not because of the inherent difficulties of outcome-oriented assessments. Three assumptions undergird the approach and findings in this report. In our view, security cooperation assessments should be important for the Air Force because

1. Assessment benefits the things assessed.
2. Air Force security cooperation can benefit if there is a common understanding of security cooperation assessments.
3. Security cooperation assessments can support DoD and Air Force decisionmaking if they are available, current, accurate, and configured appropriately.

The first assumption can be validated by Air Force examples from realms other than security cooperation. For instance, the Flag Measurement and Debriefing System has led to improvement in Air Force aerial combat proficiency; Nuclear Surety Inspections have done the same for individual and unit proficiency in nuclear weapons–related tasks and operations; and Operational Test and Evaluation improves Air Force acquisition practices and their ability to satisfy the needs of warfighters. Although assessment is generally beneficial, we acknowledge that this may not be true in all cases: for example, when the thing being assessed may not be measurable, when assessment may be prohibitively expensive, or when assessment is likely to have no effect on decisionmaking.

This report also examines various aspects inherent in the second and third assumptions by leveraging RAND's considerable experience working with the Air Force on issues related to security cooperation, specifically the 2008 effort, which created a framework for assessments, in addition to drawing on the information sources detailed earlier. The large questions at issue are as follows:

- What can a focused, integrated security cooperation assessment system do for the Air Force?
- What kinds of decisionmaking could security cooperation assessments really inform?
- How would this information benefit the Air Force and its broader force employment and force development decisions?
- How big an effort would it be for the Air Force to initiate a comprehensive security cooperation assessment program?

More specifically, this report will return to the RAND assessment framework to consider how it can best support decisionmaking and aid the Air Force in understanding what would be involved in using assessments to inform its management practices. This report will also identify specific ways in which security cooperation assessments could be conducted to best inform Air Force decisionmaking, describe the Air Force's current capacity to perform the appropriate assessments, and acknowledge the additional commitments in terms of manpower, time, training, and budget that would be necessary to create an improved system of security cooperation assessments.

Defining Key Terminology

Key terms that are used throughout this report require explanation up front. Security cooperation and its subset, security assistance, have a long history. According to the Defense Security Cooperation Agency, *security cooperation* includes

> those activities conducted with allies and friendly nations to: build relationships that promote specified U.S. interests, build allied and friendly nation capabilities for self-defense and coalition operations, (and) provide U.S. forces with peacetime and contingency access.[8]

---

[8]   See the Defense Security Cooperation Agency website's Frequently Asked Questions section. For a complete discussion of Title 10 and Title 22 authorities, as well as a general description of security cooperation and security assistance, see the Defense Institute of Security Assistance Management, DISAM's Online Green Book, 2007.

Examples include training and combined exercises, operational meetings, contacts and exchanges, security assistance, medical and engineering team engagements, cooperative development, acquisition and technical interchanges, and scientific and technology collaboration.[9]

Security assistance is a subset of security cooperation and consists of "a group of programs, authorized by law that allows the transfer of military articles and services to friendly foreign governments."[10] Examples of these programs include Foreign Military Sales, Foreign Military Financing, International Military Education and Training, and Direct Commercial Sales.

While not yet codified in Air Force doctrine, the Air Force has adopted the Joint Capabilities Area concept of *building partnerships*, defined as

> The ability to set the conditions for interaction with partner, competitor or adversary leaders, military forces or relevant populations by developing and presenting information and conducting activities to affect their perceptions, will, behavior, and capabilities.[11]

Building Partnerships includes two integral elements—communication and shaping—that is, using words and non-combat operations to achieve strategic objectives. Although sometimes used as such, *building partnerships* is not a synonym for *security cooperation*. Rather, it refers to the capabilities required for DoD to engage in peacetime engagement activities with a range of domestic and international partners.

In the 2006 *Building Partnership Capacity Roadmap*, *building partnership capacity* is defined as "targeted efforts to improve the collective capabilities and performance of the Department of Defense and its partners."[12] The 2006 *Quadrennial Defense Review* emphasizes efforts to build the security and defense capabilities of partner countries so that they can make valuable contributions to coalition operations and improve their own indigenous capabilities.[13] *Building partnership capacity* should not be used interchangeably with *building partnerships*. The former is a component of the shaping objective that building partnerships capabilities are designed, in part, to provide.

Other key terms used throughout the report include *funding source, initiative, program, activity,* and *event. Funding sources* are large umbrella resource streams that fund initiatives or programs. The Freedom Support Act, which authorizes resources for many initiatives and programs in Eurasia, is an example of a funding source. The Freedom Support Act authorizes funding, for example, for the State Department's Export Control and Related Border Security program.

---

[9]   Jennifer D. P. Moroney, Kim Cragin, Eric Gons, Beth Grill, John E. Peters, and Rachel M. Swanger, *International Cooperation with Partner Air Forces*, Santa Monica, Calif.: RAND Corporation, MG-790-AF, 2009.

[10]   U.S. Department of Defense, *Security Assistance Management Manual*, DoD 5105.38-M, October 3, 2003. A full list of security assistance programs may be found on p. 33 of the manual.

[11]   Approved Joint Capabilities Area Lexicon, January 2009, and codified within the Quadrennial Roles and Missions Review.

[12]   U.S. Department of Defense, *Building Partnership Capacity: QDR Execution Roadmap,* Washington, D.C., May 22, 2006b, p. 4. This report is an evolving document. It not only includes guidance on how DoD should train and equip foreign military forces, but also discusses the need to improve the capacity of other security services (i.e., stability police, border guards, customs, etc.) within partner countries. Moreover, the concept also refers to the need to improve DoD's ability to work with nonmilitary forces (i.e., U.S. interagency, nongovernmental organizations, coalition partners, and the private sector) in an operational context for integrated operations.

[13]   U.S. Department of Defense, *Quadrennial Defense Review Report*, Washington, D.C., February 6, 2006a; DoD, 2006b.

Initiatives are funding sources for a collection of programs that pursue a particular set of goals. An example of an initiative is the Warsaw Initiative Fund, which funds programs in central and southern Europe as well as Eurasia. The Warsaw Initiative Fund supports some Air Force security cooperation activities, such as Regional Airspace Initiative studies that have taken place in Eastern Europe.

Programs are sets of activities or events that are coordinated to achieve a certain set of objectives. Although some security cooperation programs executed by the COCOMs and component commands rely on multiple sources of funding, this analysis concentrates on those programs that are managed by the Air Force, such as OET and MPEP, which RAND addressed in its survey (see Chapter Four).

Activities and *events* are actions directed, funded, and/or supervised by program managers. Activities are particular kinds of interactions funded by programs, such as defense and military contacts (e.g., Air Force OET program). In contrast, events are specific, scheduled, time-delimited interactions that incorporate U.S. and partner representatives (e.g., specific U.S.-Chile OET meetings).

Organization of This Report

This report is organized in the following way. Chapter Two first summarizes RAND's 2008 conceptual framework for assessing Air Force efforts to build partnerships with foreign countries.[14] It then focuses on the RAND "Assessment Day" by describing and analyzing the insights, suggestions, and concerns of nearly 40 Air Force security cooperation stakeholders who participated in the discussions with respect to assessment in general and RAND's assessment framework in particular. Additionally, the chapter addresses presentations from the Air Force Campaign Support Plan Working Group's plenary sessions in cases where those remarks have bearing on the suitability of the security cooperation assessment framework. The chapter also includes the research team's discussions with former Secretary of the Air Force Michael Wynne, with whom the team spent several hours before the working group convened.

Chapter Three begins by describing the analytical approach RAND used in a targeted survey tool to gather and understand information regarding Air Force security cooperation assessments. Next, it highlights a set of overall findings that appear to be common across various programs, including insights from two specific programs: OET and MPEP. The chapter concludes by offering a discussion of the implications for Air Force security cooperation assessments.

Chapter Four addresses four basic security cooperation assessment issues:

- Why assessments should be done
- What should be assessed
- How the assessments ought to be conducted
- Who might perform them.

Chapter Five returns to the security cooperation assessment assumptions that were presented in this introduction and summarizes our specific findings and recommendations.

---

[14] Moroney et al., 2010a.

Appendix A describes the approach used in developing our assessment survey of Air Force security cooperation program stakeholders, provides a listing of the questions used to elicit responses regarding assessment roles and data types, and provides a series of tables that show the results of the survey using the analytical construct.

Appendix B provides a generic version of the RAND assessment survey taken by Air Force security cooperation stakeholders.

CHAPTER TWO

Air Force Perspectives on Security Cooperation Assessments

More often than not, those attempting to assess security cooperation activities tend to zero in on measures or criteria to enable the assessment, rather than taking the broader approach of an actual assessment framework that includes a spectrum of measures, players, and structures to enable a more comprehensive assessment. RAND's 2008 security cooperation assessment framework exemplified a broader approach. As noted in Chapter One, however, a diversity of opinion exists within the Air Force on many aspects of security cooperation assessments. Consequently, successful implementation of a comprehensive approach requires that Air Force leaders take into account the variety of stakeholder views regarding assessment both to fine-tune the approach and enhance its attractiveness and utility for security cooperation practitioners.

This chapter describes RAND's original security cooperation assessment framework[1] and reports reactions to it gathered during an "Assessment Day" workshop attended by Air Force officers and civilian officials. It also summarizes the broader discussions of security cooperation assessment issues from that workshop.

Assessment Framework

RAND's 2008 security cooperation assessment framework included the following five elements:

1. Ends: the objectives sought by security cooperation activities, which are typically expressed in Air Force guidance
2. Ways (activities and programs) and means (resources and authorities) through which the Air Force achieves its ends
3. Stakeholders and their roles
4. Hierarchy of evaluation
5. Measurement categories and data.

Each of these five elements will be described in the following sections.

Ends or Objectives
Guidance for security cooperation assessments comes from a number of sources. Based on the strategic guidance, the Office of the Secretary of Defense (OSD) produces the Guidance for

---

[1] For a more complete description, see Moroney et al., 2010a.

Employment of the Force,[2] which COCOMs use as the basis for developing their theater campaign plans. The Guidance for Employment of the Force includes focus areas that are intended to help the COCOMs, services, and defense agencies concentrate their security cooperation efforts with partner countries. However, the focus areas covered in the Guidance for Employment of the Force are quite general. The Guidance for Development of the Force[3] is meant to be more directly relevant to force development, and thus is useful to the Air Force in its force provider role.

At the Air Force level, the 2008 Air Force Global Partnership Strategy makes building partnerships the centerpiece of Air Force security cooperation activities. Described as a critical link between national-level strategy and Air Force Campaign Support Plans, this Air Force document establishes the following end states:

- Mutually beneficial global partnerships are established, sustained, and expanded
- Global partners have the capabilities and capacity necessary to provide for their own national security
- The capacity to train, advise, and assist foreign air forces, and conduct security cooperation activities, using airmen with the appropriate language and cultural skills, is established
- Partnership interoperability, integration, and interdependence are developed and enhanced.

Ways or Activities and Programs

Air Force security cooperation tools are often characterized as "ways" and typically represent categories of activities that the Air Force undertakes with a partner air force in pursuit of Air Force security cooperation goals, or "ends." Thus, there emerges a means-ways-ends relationship in which "means"—funding, personnel, and other resources—are organized by "ways," such as education or exercises, in pursuit of "ends," or goals.

It has been an accepted practice for major security cooperation organizations (e.g., OSD, the services, COCOMs, and the Defense Security Cooperation Agency) to develop their own list of security cooperation ways, which they periodically modify. Fortunately, most of these lists share many elements in common. The following security cooperation ways appear in the 2008 Guidance for Employment of the Force:

- combined and multinational education
- combined and multinational exercises
- combined and multinational experimentation
- combined and multinational training
- counternarcotics assistance
- counter- and nonproliferation
- defense and military contacts
- defense support to public diplomacy
- facilities and infrastructure support projects

---

[2]   Office of the Secretary of Defense, *Guidance for Employment of the Force*, Washington, D.C., July 18, 2008; not available to the general public.

[3]   U.S. Department of Defense, *Guidance for Development of the Force: Fiscal Years 2010–2015*, Washington, D.C., April 2008; not available to the general public.

- humanitarian assistance
- information sharing and intelligence cooperation
- international armaments cooperation
- security assistance
- other programs and activities.

Stakeholders and Their Roles

So, too, there are many assessment stakeholders. Stakeholders are those organizations or persons with a role in planning, resourcing, or executing the various security cooperation activities or programs. Key Air Force stakeholders include, but are hardly limited to,

- SAF/IA
- AF/A3/5
- Air Force component commands
- other major Air Force commands.

At each level of assessment, there are five functional assessment roles that the Air Force and other agencies perform with respect to security cooperation programs, whether they are managed by the services, by the COCOMs or OSD, or by the State Department. In some instances, these roles are clearly spelled out in laws, policy directives, and program instructions. In other cases, they must be inferred by taking into account the character of the organization and the extent of its de jure and de facto decisionmaking authority. The functions of these roles are data collector, assessor, validator, integrator, and recommender. Data collectors gather information to support a specific assessment. Assessors conduct the assessment. Validators review the assessment for its objectivity and completeness. Integrators organize and synthesize assessments to meet OSD and Air Force requirements. Recommenders formulate specific recommendations for the chain of command based upon the results of the assessment. In the interest of objectivity, the data collector and assessor roles should generally be separated from the validator, integrator, and recommender roles.

There are, indeed, notable differences in the roles that various stakeholders play. Plainly, there are both Air Force service and joint stakeholders, and the specific stakeholders vary between those with a program focus and those with a country focus. It also matters whether the activities in question are managed by the Air Force, like OET. Some stakeholders are responsible for playing a direct role in the program assessment process, while others are consumers of the assessments themselves—such as the Office of Management and Budget and Congress. If the program has an accompanying directive that specifies the responsibilities of the various stakeholders, this makes assigning assessment roles all the easier.

Hierarchy of Evaluation

The assessment framework serves to sustain the explicit focus on assessment for decisionmaking, and to connect stakeholders to specific types of assessment for their decision needs. This framework is based on the "hierarchy of evaluation" presented in Figure 2.1, which includes five levels of assessment for purposes of the security cooperation program assessment process.

The hierarchy of evaluation should be applied as a continuous cycle, starting with Level 1 at the bottom, continuing through each assessment up to Level 5, and then returning to the beginning. This is an iterative cycle in which, logically, senior leaders should be responsible for determining the answers to both Level 1—Is there a need?—and Level 5—Does this program

Figure 2.1
Hierarchy of Evaluation

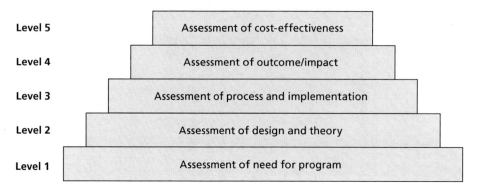

Level 5	Assessment of cost-effectiveness
Level 4	Assessment of outcome/impact
Level 3	Assessment of process and implementation
Level 2	Assessment of design and theory
Level 1	Assessment of need for program

SOURCE: Adapted from Figure 7.1 in Christopher Paul, Harry J. Thie, Elaine Reardon, Deanna Weber Prine, and Laurence Smallman, *Implementing and Evaluating an Innovative Approach to Simulation Training Acquisitions*, Santa Monica, Calif.: RAND Corporation, MG-442-OSD, 2006, p. 110.
RAND *TR907-2.1*

do better than others from a comparative cost-effectiveness perspective?—as their judgments about relative cost-effectiveness at Level 5 would, in turn, drive decisions about which needs at Level 1 were still important and which were no longer so.

At each level, there are specific assessment measures and metrics. For instance, Level 1 assessments focus on the problem to be solved or goal to be met, the population to be served, and the kinds of services that might contribute to a solution. Assessment questions at this level would include the following:

- What are the nature and magnitudes of the problems to be addressed?
- What audience, population, or targets does the need apply to?
- What existing programs or activities contribute to meeting this goal or mitigating this problem?
- What are the goals and objectives to be met through the policy or program?

Once a need for the program is established, one can proceed through the remaining steps of the hierarchy of evaluation. Level 2 asks about concept, design, and theory in connecting activities to the strategic goals that spawned the needs. Level 3 asks how well the design is being implemented. It includes "outputs," the countable deliverables of an activity or program, and it is the place for a limited assessment of efficiency: Is implementation getting the most possible out of the available resources? Level 4 turns the attention from outputs to outcomes and their impact. If *outputs* are the immediate products of activities, then *outcomes* are the more enduring changes resulting from those activities. Since they are meant to be enduring, outcomes are often very difficult to measure. Finally, Level 5 assesses relative cost-effectiveness—"bang for the buck."

Assessment Measures

To effectively gauge the results of security cooperation efforts, it is necessary for the Air Force to have specific assessment measures that are consistent over time, as well as appropriate metrics that demonstrate levels of commitment, progress, and achievements, which translate into the three measures in the RAND assessment framework:

- **Inputs** are commitments and resources, including money, manpower, skills, and materiel, that are required to execute an event.
- **Outputs** show progress—the direct products of an event, activity, or program, e.g., numbers of graduates, items delivered, or crews certified.
- **Outcomes** are achievements that can be thought of as the effect of one or more outputs on the target audience (e.g., a favorable view of the United States), or changes in program participants' behavior (e.g., new receptivity to U.S. Air Force advice), knowledge, skills, status, and/or level of functioning.

Air Force Reactions to the Assessment Framework and Views on Assessment

To engender frank and open discussion of RAND's proposed security cooperation assessment approach, the research team held an "Assessment Day" workshop on May 15, 2009, gathering more than 40 participants representing organizations from across the Air Force that are engaged in Air Force security cooperation activities. Using the assessment approach, including the hierarchy of evaluation and the various assessment roles as a basis for discussion, the study team sought to acquire insight into how and whether assessments are currently taking place, gain feedback on the RAND assessment approach, and better understand factors that might hinder its implementation.

The Assessment Day workshop was held as a follow-up meeting to the Air Force's Campaign Support Plan Working Group, which was hosted at RAND earlier in the week, and afforded us an opportunity to interact with a representative cross-section of security cooperation stakeholder representatives.[4] All had experience with, or an active interest in, Air Force security cooperation assessment efforts. Because this group was self-selected from a group already attending the Campaign Support Plan Working Group, it represented a wide range of stakeholder perspectives and interests, and included representatives from specific program offices, those with regional (or even global) portfolios, and those with interest in specific countries.

Assessment Day began with a brief plenary in which the core components of RAND's proposed assessment framework were presented: the hierarchy of evaluation, the proposed assessment roles, and the proposed supporting resources. Following the plenary, participants were divided into working groups of nine to eleven persons for an extended working session. Each of the four working groups included two members of the RAND team: one as facilitator and the other as rapporteur.

Each facilitator led a free-form discussion loosely structured around an agenda common to each working group. The emphasis in all groups was on understanding stakeholder perceptions of the proposed assessment framework and frank discussion of its practicality. The agenda

---

[4] Participants included representatives from Headquarters U.S. Air Force (HAF) International Affairs; Intelligence, Surveillance and Reconnaissance (AF/A2); Logistics, Installations, and Mission Support (AF/A4/7); Operations, Plans and Requirements (AF/A3/5); Analyses, Assessments and Lessons Learned (AF/A9); 13th Air Force; 17th Air Force; Air Forces Central; Air Combat Command; Air Education and Training Command (AETC); Air Force Special Operations Command; Air Mobility Command; U.S. Air Forces in Europe; the Coalition and Irregular Warfare Center of Excellence; Pacific Air Forces; and the Air Force Research Institute.

included a discussion of the proposed assessment roles and attempts to identify specific offices that currently perform, or could perform, these roles in the future; which specific offices contribute, or should contribute, assessments and at which levels of the hierarchy of evaluation; and which elements to support assessment are currently available in the Air Force security cooperation community and to what extent.

In the remainder of this chapter, we first explain how we analyzed the participants' comments and insights, then we summarize the discussions within the breakout groups.

Analytical Approach

We employed two analytical tools to help us identify and understand the main themes emerging from the Assessment Day breakout groups. The primary tool we used was text analysis software called Atlas.ti, which is often used to help researchers understand large volumes of text, imagery, and other media.[5] It brings rigor to the process of identifying common themes and points of view, and it allowed us to determine the degree of sameness ("congruence") among the four individual breakout groups. The software also supports "mapping" through the breakout session rapporteurs' notes in order to identify and integrate threads of conversation linking the various elements of the security cooperation assessment framework, determining whether the participants think the Air Force has the necessary skills, authorities, raw information about security cooperation activities, and personnel to fill the necessary assessment stakeholder roles as proposed in RAND's assessment framework. We employed another software application, QDAMiner, as an independent control on Atlas.ti, and to build graphics to display some of the data.

Congruence

The notion of congruence among the opinions about the security cooperation assessment framework captured in the rapporteurs' notes was important because of the small number of participants in the breakout groups. If the individual groups reflected very different perspectives on the security cooperation assessment framework, it would be difficult to generalize from their views. On the other hand, the more congruence reflected in the notes, the greater the consensus among the participants on the vices and virtues of the security cooperation assessment framework, and the greater the likelihood that their views reflect those of the larger security cooperation community.

Mapping

To "map" the most significant common issues of concern among the Assessment Day participants, we looked for evidence of all the essential ingredients for the security cooperation assessment framework that could be found in the notes taken during the breakout groups. We sought to discover whether our Assessment Day participants believed each level in the hierarchy of evaluation would be feasible in the context of security cooperation programs, country security cooperation portfolios, and security cooperation activities and events. We also looked for participant feedback on the proposed assessment stakeholder roles—specifically, whether our participants could identify potential data collectors, assessors, assessment validators, and

---

5   Developed at the University of Berlin, Atlas.ti is now in its seventh edition.

integrators within their own organizations or elsewhere in the chain of command. Finally, we looked for evidence in the breakout group discussions that indicated whether the Air Force had sufficient resources—specifically time, personnel, skills, guidance, and data—to implement the security cooperation assessment framework.

Assessment Day Summary

This section summarizes the views from the four breakout groups on the hierarchy of evaluation, the stakeholder roles, and the sufficiency of key assets to support assessment. The chapter closes with the research team's conclusions from the Assessment Day activities.

Hierarchy of Evaluation

RAND's proposed hierarchy of evaluation received mixed reviews among breakout group members. In part, these views reflected misperceptions about the hierarchy and its constituent assessments, which are described below.[6] However, some concerns raised reflected legitimate worries. Some of the participants stumbled on Level 1 of the hierarchy—assessment of the need for a program—preferring country-oriented rather than program-oriented assessments, even though the hierarchy could easily be adapted for application either way. In our quantitative analysis of the Assessment Day discussions, 21 percent of the participants expressed concern that the objectives of security cooperation assessment were unclear.

Other Assessment Day participants voiced a number of concerns. They argued that the security cooperation assessment framework would

- miss important qualitative factors
- overemphasize near-term, quantitative elements that would distort the near-term value assigned to security cooperation
- miss the longer-term effects and the indirect, second-order benefits altogether.

Still others raised measurement issues: How would the security cooperation assessment framework manage cost-effectiveness across countries? What about program outcomes, as opposed to outputs? At least one participant voiced concern that assessment is a joint venture and that the security cooperation assessment framework did not capture that fact.

In addition, 21 percent of participants cited process difficulties in attempting security cooperation assessment and considered the lexicon and doctrine associated with security cooperation as obstacles to assessment.

The feedback we received from the Air Force security cooperation community highlighted the need for us to strengthen the case for a comprehensive approach to assessment. However, many of the specific objections raised to RAND's assessment framework appear unwarranted. First, there is nothing in our framework that excludes the consideration of qualitative factors in an evaluation of Air Force security cooperation efforts. Indeed, we believe qualitative data would be especially useful for conducting outcome/impact assessments. Second, we are not proposing that the Air Force focus its security cooperation assessments on near-term results. In

---

[6]   Research team members in the breakout groups made limited attempts to correct misapprehensions, preferring instead to reserve the time for a complete discussion of the security cooperation assessment framework.

the case of outcome and cost-effectiveness assessments, for example, we foresee security cooperation data being collected over a period of several years and studied using time-series analysis. Third, we do not think that security cooperation assessments will inevitably neglect the indirect or delayed effects of security cooperation activities, given that many of these are known to airmen with security cooperation experience. Finally, the concern about the lack of "jointness" in our framework can be easily countered in that we specifically call for the engagement of security cooperation experts from the other services in the assessment process. That said, the fact that a significant number of Assessment Day participants raised objections to RAND's assessment framework may indicate that it should be simplified so that it can be more easily comprehended by those who might be tasked to use it.

Not all of the points raised by Air Force security cooperation officials were negative with respect to assessment in general or RAND's assessment framework in particular. Forty-two percent of the Assessment Day participants expressed faith in the proposition that long-term, qualitative assessments would have benefits for Air Force security cooperation.[7] Some participants also found RAND's hierarchy of evaluation quite comprehensive, and several claimed to have performed all levels of assessment themselves. One of the team's facilitators managed to demonstrate to his breakout group the full versatility and flexibility of the hierarchy, as well as its ability to accommodate every example they posed.

Assessment Roles

The breakout groups generated more feedback about the stakeholder roles than they did about the hierarchy of evaluation. Some participants indicated they could identify all of the assessment roles within their organizations and chains of command. Several other participants self-identified with the role of data collectors. Others suggested that there might be a need for an additional role, that of "synchronizer" between the assessment process and the decisionmakers who ultimately consume the assessments. Although the integrator role already identified can probably handle the suggested synchronization functions, the team took this suggestion under advisement.

Based on our quantitative analysis of the Assessment Day discussions, 35 percent of participants accepted the basic conception of stakeholder roles. Twenty-eight percent had some dispute with or misgivings about stakeholder roles. The other 37 percent had no opinion on this topic.

One of the presentations in the plenary discussion asserted that "aggregation" of data (as opposed to actual analysis) was the limit of the Air Force's capabilities today. If the assertion is correct, it would indicate that the Air Force may have adequate numbers of data collectors but lack potential assessors, validators, integrators, and recommendation formulators.

Key Assets to Support Assessment

Discussions of the key assets that are needed to conduct assessments provoked the most discussion among the Assessment Day participants. Their concerns focused principally on guidance, access to data and information, skills necessary to perform assigned roles, authorities (for access to information, to establish supporting-supported relationships, and to specify assess-

---

[7] The remaining 58 percent of Assessment Day participants did not express an opinion on this topic. However, as indicated above, some individuals were skeptical about the Air Force's ability to carry out long-term, qualitative assessments, either with or without the use of RAND's framework.

ment instructions, data flows, and procedures), and adequate resources (people, time, skills, money). Each of these concerns will be discussed below.

Guidance. The abundance of disparate assumptions about security cooperation assessments among the participants, including what to assess, for what reasons, how to assess, and who might conduct assessments, suggests a clear need for guidance. Ideally, the Air Force should first identify the decisions that would benefit from assessment inputs and then specify the supporting assessments. Some of the participants in the breakout groups thought that greater coherence across existing guidance was needed concerning goals, objectives, instructions, and specific criteria to be satisfied. There was a broad sense among the participants that guidance specific to assessments is needed about objectives, metrics, information flows, and critical data. Indeed, 57 percent of participants commented on the need for guidance in order to perform appropriate assessments (this issue had the highest percentage, based on our quantitative analysis of the issues discussed). Without guidance, Air Force officers attempting assessments have a variety of possible options available, which could produce widely varied outcomes. To avoid this outcome, if the Air Force concludes that it needs a security cooperation assessment, it should focus on designing the assessment process, writing instructions for those who must conduct assessments, identifying the participants and their roles, and specifying the information necessary to conduct the assessment.

Access to Data. Assessment Day participants widely identified access to data and information as an impediment to security cooperation assessment. According to one view, collaboration and data-sharing habits are not fully established. Another perspective posited that data are abundant but hard to get because the sources of relevant information are not intuitive. Several participants noted that the offices that have relevant information are not necessarily those who need it, and they may not be aware of other offices that could benefit from their data. Another participant noted there is no established process for moving information to the people and offices that need it. Based on our analysis, 42 percent of participants indicated there was a greater need for information sharing in order to make assessment possible, and 35 percent expressed concern about the difficulty associated with getting data to support security cooperation assessments. In addition, some expressed concerns that quantitative data can bias assessments toward the near term. This could be the case unless organizations retain near-term data over time and aggregate such data so that they are available for longer-term trend analysis. Another participant was concerned that current Air Force security cooperation data do not reflect qualitative factors.

Necessary Skills. Assessment skills were widely cited as a shortfall. Although some participants were very confident in their ability to perform certain kinds of assessments, particularly needs assessments, others expressed their doubts about Air Force personnel being prepared to handle all levels of assessment as reflected in the hierarchy. Indeed, 42 percent of participants said that their organizations had limited skills to conduct assessments.[8] One observed that professional military education does not prepare officers for the full hierarchy of the evaluation menu. Indeed, many of the participants remarked on the need for specific training. Several

---

[8]   Although a few individuals claimed that their organizations did possess the necessary skills to perform security cooperation assessments, the majority of Assessment Day participants did not express an opinion on this topic, either because they did not know or did not wish to present their views. We took another look at the skills issue in our survey of Air Force officials involved in the MPEP and OET programs. The responses to the survey presented a more complex picture of the skills issue than did the Assessment Day focus groups. See Chapter Three.

breakout group members remarked pointedly that operations research alone was an inadequate skill for conducting security cooperation assessments, and that other skills affording an appropriate appreciation of qualitative factors were necessary to render sound assessments within the security cooperation realm. They felt that the qualitative dimension of assessments had to be more directly addressed. One officer offered Eagle Look as a model for how the Air Force might overcome skill impediments, although the research team doubts that a comprehensive review of security cooperation would be necessary on a regular basis.[9]

Authorities. Relatively little was said about requisite authority for security cooperation assessments. One group identified authority as a "big problem," noting there are substantial obstacles to achieving full cooperation and information access among busy organizations without clear authorities that would compel cooperation and data sharing. Despite this concern, most participants believed they had adequate authority to do security cooperation assessments, although there was discussion that a lack of authorities could lead to country-specific issues, organizational problems, and a need for the synchronization of efforts.

Resources. Although most did not express an opinion, a sizable group of participants viewed resources as a serious constraint: 28 percent indicated that their organizations lacked the time, personnel, funding, and skills with which to conduct assessments. Participants often remarked on the significant amount of assessment that is already ongoing and questioned the value of ordering additional security cooperation assessments. This line of discussion seemed to resonate among those who were skeptical of assessments' ability to capture long-term and indirect effects of programs and among those who felt that the qualitative dimension of assessments is important and undervalued in the Air Force.

Conclusions

The Assessment Day breakout groups provided rich insights into expert views on the security cooperation assessment framework and its utility for the Air Force. Several observations stand out.

First, it is not clear that all of today's Air Force security cooperation assessments are decision-oriented. Some, typically process-implementation assessments, clearly are decision-oriented, insofar as they contribute to the Air Force's ability to decide the degree to which its activities are operating in accordance with their instructions and directives. Because many of the discussions in the breakout groups treated "assessment" as a means of determining the suitability of a given country to be a partner, the participants considered these individual assessments to be decision-oriented (i.e., should the Air Force engage country X in security cooperation or not?). More broadly, however, among Assessment Day participants, there was no clear indication of specific decisions that security cooperation assessments are intended to support. Rather, participants perceived assessments to be useful for just figuring out "what's going on" and telling senior officials about it in a timely manner.

Second, the Assessment Day participants indicated, the Air Force is not ready for assessments as conceived by the security cooperation assessment framework. They identified a

---

[9]   The Eagle Look Directorate within the Air Force Inspection Agency conducts Eagle Looks, or management assessments, of Air Force operations, support, logistics, maintenance, and acquisitions processes and programs. See USAF Fact Sheet, "Air Force Inspection Agency," March 2009.

number of gaps that the Air Force must fill in order to conduct adequate security cooperation assessments. The first thing needed is guidance: what to assess and how to assess it; what the metrics for assessment will be and where the data will come from; how the data will be configured; the counting rules to be used; and the frequency at which the assessment will take place. Participants felt that guidance also must address the stakeholder roles and identify the actual actors who will fill them.

The next requirement they identified is formal training. Designing and conducting assessments is not necessarily intuitive. Therefore the Air Force, if it determines it wants more security cooperation assessments than it currently undertakes, will have to specify and train personnel appropriately to perform all of the stakeholder roles.

Finally, a widely held view among the RAND Assessment Day participants was the need for sufficient personnel to support assessments. If current Air Force security cooperation officials have their hands full with their present responsibilities, additional assessments would seem to require more manpower. However, a "smart" assessment mechanism that demanded only the minimum information necessary for decisionmaking purposes—and established clear data requirements, reporting relationships, and assessment roles—would not necessarily increase the personnel requirements of the existing system of Air Force security cooperation management and execution. But the near- and long-term costs of developing such a mechanism depend on the Air Force's present capacity for security cooperation assessment and any changes that might be needed to improve that capacity; this topic is explored in Chapter Five.

Not all of the uncertainties attending security cooperation assessment are up to the Air Force to resolve. Some assessment issues can only be tackled through a collaborative effort involving OSD, the Joint Staff, the COCOMs, and other military departments and defense agencies, such as the Defense Security Cooperation Agency. If the Air Force is to provide useful security cooperation assessments, OSD and the Joint Staff, in particular, should be clear about the decisions the assessments are meant to inform. The following are a number of key issues that could be addressed via a joint security cooperation approach:

- At what level within the hierarchy of evaluation should assessments be conducted?
- Should the assessments be country- or program-oriented?
- Should the assessments emphasize qualitative or quantitative measures?
- How frequently should assessments take place?
- What coordinating instructions are available to help organize the efforts of individual participating organizations?

Within the Air Force, SAF/IA and AF/A3/5 should take the lead in raising these fundamental assessment issues and helping to develop an Air Force and, eventually, a DoD consensus that could be disseminated through high-level documents like the Campaign Support Plan and the Guidance for Employment of the Force. Venues for such consensus-building discussions could include the annual Air Force Building Partnerships Conference and the Air Force Campaign Support Plan Conference, to which selected DoD security cooperation officials outside the Air Force might be invited, as well the COCOMs' annual Theater Security Cooperation Conferences, in which a number of Air Force security cooperation organizations participate.

In the interim, the RAND study team has begun to tackle some of the obstacles to implementing a comprehensive approach to security cooperation assessments that were identified during Assessment Day and other discussions with Air Force officials. In Chapter Three, we focus on the issue of the Air Force's capacity to conduct a variety of integrated assessments within the context of two Air Force–managed programs. In Chapter Four, we try to address the skepticism evident within the Air Force security cooperation community regarding RAND's comprehensive assessment concept. We do this by explaining the importance of assessments in making high-level decisions on security cooperation programs and resources, outlining the basic elements of a future assessment guidance document, and illustrating how these elements might be applied to the two programs highlighted in Chapter Three.

Understanding the Air Force's Current Capacity to Conduct Security Cooperation Assessments

Our understanding of the Air Force's current capacity to conduct a broad range of security cooperation assessments is complicated by the fact that every Air Force security cooperation program has multiple stakeholders, often scattered across the Air Force at different levels and with different priorities, perspectives, and capabilities. Joint stakeholders may also have information about Air Force security cooperation programs, and sometimes they even participate in activities or provide other resources that contribute to the program. To understand the roles these various stakeholders play in these programs, and, more importantly, what roles they play or could play in program assessments, the RAND study team collected information from Air Force and joint stakeholder representatives using two separate methods.

This chapter begins by describing the analytical approach RAND used to gather and understand information regarding Air Force security cooperation assessments. Next, it highlights a set of overall findings that appear to be common across various programs, including insights from two specific programs: MPEP and OET.[1] The chapter concludes by offering a discussion of the implications for Air Force security cooperation assessments.

Survey of Air Force Security Cooperation Experts

The team used two separate approaches to ascertain the extent to which the security cooperation assessment approach detailed in its previous work is feasible for employment by the Air Force security cooperation community. The first was the RAND Assessment Day workshop described in Chapter Two. Second, the team gathered quantitative supporting analyses through aggregate responses to a structured survey, titled the "Project AIR FORCE Air Force Security Cooperation Request for Expert Feedback."

Constructing the Survey

Developing a survey that could elicit data from real stakeholders required the study team to move from the abstract nature of the evaluation hierarchy to a more concrete one that describes its essential parts and their relationships. To do this, the study team asked questions regarding current program tasks, as opposed to theoretical assessment tasks. In addition, the survey drew heavily on existing Air Force security cooperation program design and development guidance as a way to ensure that the terminology in it was familiar to the respondents.

---

[1] OET was previously called the Operator-to-Operator Program or Operator-to-Operator Staff Talks.

The survey questions were grouped by four broad areas that a stakeholder might be involved in with respect to a program:

- **Process Implementation.** Some stakeholders carry out specific tasks as assigned by program managers. These tasks might include organizing an event or providing subject-matter expertise, establishing contracts, accounting for funds, or processing documentation required by Air Force instructions (AFIs) or other directives.
- **Process Design and Development.** Other stakeholders participate in the design or development of processes, carrying out such activities as developing lesson plans, contracts, or event agendas.
- **Make Recommendations.** Some stakeholders make recommendations to program managers about the size, scope, or need for the program or a specific activity.
- **Make Decisions.** Still other stakeholders make decisions regarding the specific activities, the need, or the scope of the program.

Using this structure to identify the various stakeholders, the study team deployed an online tool to elicit information from actual Air Force stakeholder representatives. The survey employed terms and references that were familiar to the respondents, and it asked concrete questions about activities the respondents were currently engaged in.

By mapping their answers back to assessment roles, the study team identified where potential assessment capability exists and where there are possible gaps, and the team developed insights that can inform Air Force efforts to implement a comprehensive assessment framework as described in this report.

Because of the role they play in guiding the development and conduct of Air Force activities, the study team reviewed AFIs and similar documents to identify the types of documentation and data collection required for each of the programs. Box 3.1 on the following page shows the five categories that were chosen to organize the types of data that could be collected and assessed across an entire program. Under each category, the typical documents prepared, collected, or reviewed in association with Air Force security cooperation programs are listed. These documents are specifically cited in various security cooperation–related AFIs, and were included as a way to make the survey content more immediately accessible to the respondents, i.e., to represent documents that respondents created or used in their day-to-day duties.

The study team approached the construction of these questions from the understanding that program guidance forms the basis for assessment. In other words, program guidance documents the need for a program and its objectives, and program guidance often tells program managers how to design and implement their programs. The most common type of guidance documents associated with Air Force security cooperation programs are Air Force Policy Directives (AFPDs) and AFIs. With the exception of OET, each of the Air Force security cooperation programs examined had an associated AFPD and AFI.

Air Force Guidance: Policy Directives and Instructions

While documents such as AFIs exist for most Air Force security cooperation programs, these documents provide only top-level guidance and are often supplemented by stakeholders' organizational handbooks, checklists, operating instructions, and other documents that capture local procedures and ensure continuity as programs change hands. Program managers and program stakeholders may, because of their close proximity to program activities and events,

---

Box 3.1: Types of Security Data That Can Be Collected and Typical Documents for Each Category

Demand:	Throughput:	Resources:	Cost:	Objectives:
• Country list • Country nomination • Guest list • International visit request • Invitational travel order • Letter of request • Nomination package • Program request • Project agreement/ arrangement • Project nomination form • Project proposal • Proposal for professional military education (PME) exchange • Request for Air Force personnel attendance • Request for use of Air Force aircraft • Summary statement of intent • Training quota • Travel order • Visit request	• Letter of acceptance • Country list • Country nomination • Guest list • International visit request • Invitational travel order • Letter of request • Nomination package • Request for Air Force personnel attendance • Training quota • Travel order • Visit request	• Budget allocation memo • Budget projection • Budget request • Loan agreement • Periodic financial report • Quarterly obligation report • Request for fund cite • Travel voucher	• Loan agreement • Periodic financial report • Travel voucher • Budget allocation memo • Budget projection • Budget request • Quarterly obligation report • Request for fund cite • Travel order	• After-action report • Alumni whereabouts • Annual report • Certification to Congress • End of tour report • Exchange agreement • Interim tour report • International agreement • Meeting summary/minutes • Memorandum of agreement • Memorandum of understanding • Participant entry or exit testing • Progress report • Project final report • Project quarterly report • Quarterly obligation report • Quid-pro-quo analysis • Summary statement of intent • Test and disposition report • Training report

---

have much greater insights into a program's true workings than is outlined in an AFI or other document.

This collective body of documentation—both "official" and "unofficial"—likely forms a more complete source for both data collection and assessment of Air Force activities. As such, it is important to understand what it comprises and how extensive it is.

The survey respondents were asked to comment on the availability of guidance documents, as well as to describe documents they develop themselves. Respondents were asked specifically about AFPDs, AFIs, and other program-specific documents that would guide the conduct of the program, as well as internal operating procedures.

Survey Respondents Represent a Broad Range of Stakeholders

In collaboration with Air Force officials, the study team selected two security cooperation programs as representatives of the Air Force's security cooperation efforts: the Operator Engagement Talks (OET) and the Military Personnel Exchange Program (MPEP). OET, a program managed by the HAF Regional Affairs Division, conducts a series of reciprocal staff talks between HAF and various partner air force staffs. These talks cover a range of operational issues aimed at improving the interoperability between the staffs. In contrast, MPEP, managed by the Deputy Under Secretary of the Air Force for International Affairs, is a relatively large program that conducts yearlong exchanges between USAF officers and officers from foreign air forces.[2]

Although both programs are designed to build partnerships, there are some significant differences between the two that make them useful for this study. First, MPEP is included in the Air Force's budget each year; OET leverages funding where it can, often using funding "out of hide." Second, MPEP is governed by Air Force and other formal directives that provide design and implementation guidance, whereas OET has no formal guidance pertaining specifically to it. Finally, MPEP has an established structure, with a full-time program manager and a set of regional management offices. OET is managed as an additional duty within the Office of the Deputy Chief of Staff for Operations, Plans and Requirements, Directorate of Operational Planning, Regional Affairs Division, Headquarters U.S. Air Force (AF/A5XX), and other support is provided as available upon request.

Potential respondents with expertise relevant to these programs were identified jointly with the sponsor and were then emailed an invitation to participate. The pool of potential respondents for MPEP was approximately 270 personnel, and about 60 for OET. These pools were compiled initially from applicable AFIs and were refined by SAF/IA and AF/A5XX based on their knowledge of program stakeholders and other supporting organizations.

Although the survey participants were selected based on their presumed expertise regarding one of those two programs, participants were asked to complete the survey once for each of the 16 Air Force–managed security cooperation programs about which they were knowledgeable.[3] As a result, the number of individual respondents does not equal the number of responses, because many respondents completed the survey several times.

Approximately one-quarter of the invited participants submitted surveys pertaining to OET and MPEP. Specifically, 15 responses related to OET (25 percent of those invited), while 70 were concerned with MPEP (26 percent). Both of these are statistically significant. In addition, each of the four organizational levels were represented for both programs.

Overall, respondents completed the survey a total of 149 times. Responses were not attributable to specific offices or respondents. Instead, the survey respondents were grouped according to four levels: (1) department level, (2) major command (MAJCOM)/numbered air force (NAF), (3) wing and below, and (4) joint. By grouping the respondents this way, we could maintain the respondents' privacy while gaining insight into the potential security cooperation assessment roles and an understanding of where gaps might exist.

Findings and Observations

Air Force security cooperation programs are planned, implemented, and supported by many stakeholders, each contributing to the programs' effectiveness. By analyzing the results of the

---

[2]   For a detailed description of both programs, see Moroney et al., 2010a. In that volume, OET is referred to as Operator-to-Operator Staff Talks.

[3]   Time and resource considerations led to the decision to focus the survey on OET and MPEP.

Assessment Survey, RAND was able to identify key areas that could be addressed as a way to strengthen the Air Force's ability to assess its security cooperation programs. These key areas are described below, along with the detailed examples that give insight into current gaps. See Appendix A for detailed results of the Assessment Survey.

Potential Data Collectors and Assessors Are Already in Place

The study team reviewed the responses to groups of questions that revealed information regarding assessment roles in each of the five types of assessments described in the hierarchy of evaluation. Organizing data by hierarchy level and role allowed for specific insights regarding MPEP and OET regarding the roles that stakeholder and supporting organizations might play. Moreover, this structure allowed for a way to identify gaps that might exist in data collector and assessor roles.[4] When combined with the data collected during the Assessment Day workshop, a fairly complete picture of skills, guidance, and stakeholder views on security cooperation assessments emerges.

In general, there are stakeholders across the Air Force that could contribute to assessments of Air Force security cooperation activities. These stakeholders are at the department level (i.e., HAF), MAJCOM or NAF, and at the wing level and below. In addition, joint stakeholders, such as those at COCOMs, defense attachés, and security assistance officers, contribute resources and personnel to support Air Force security cooperation programs. For example, the study team found that MPEP, based on information contained in AFI 16-107, *Military Personnel Exchange Program (MPEP)*, and our discussions with program managers, comprises several dozen stakeholders across the Air Force and outside the Air Force. In each case, one or more stakeholder respondents answered one or more questions in a way that indicated that their organization could serve as either a data collector or assessor.[5]

Table 3.1 depicts the percentage of respondents that answered at least one question positively for the data collector or assessor roles and the type of data. The table is arranged according to roles, either data collector or assessor, and also by the type of data collected or assessed. The survey questions were grouped according to the role and type of data to which they correspond (as introduced in Box 3.1).

Although at first glance it may appear that "everyone is doing everything" with regard to collecting data, the data being collected by one stakeholder are very likely different in character from those being collected by other stakeholders.

The survey results provide insights into the types of data that specific stakeholders can collect or assess. Based on the respondents' answers, in most cases it was possible to identify stakeholders who could collect specific types of data, such as demand or resources expended. We also could determine whether there is a corresponding stakeholder who could use those data to conduct an assessment. The next section illustrates this type of analysis by examining in detail how the data on objectives revealed specific insights for conducting outcome and impact assessments.

---

[4] The survey was designed to identify potential data collectors and assessors. Other assessment roles require essentially the same skills as the assessor, and decisions regarding how roles other than data collector would likely be based on chain of command or similar considerations not addressed in the survey.

[5] Although the survey was designed only to identify potential data collectors and assessors, there are additional assessment roles, including validator, integrator, and recommender. These three roles would most likely be linked to management and chain of command, and can be broadly construed to require the same attributes as those of an assessor.

Table 3.1
Positive Responses for Data Collection and Assessment Roles

Assessment Area	Level	OET		MPEP	
		Data Collector	Assessor	Data Collector	Assessor
Demand	Department	43	43	100	100
	MAJCOM or NAF	50	75	75	25
	Wing and below	n/a	n/a	56	68
	Non–Air Force	75	75	86	43
Resources	Department	43	43	100	100
	MAJCOM or NAF	50	75	75	72
	Wing and below	n/a	n/a	20	30
	Non–Air Force	25	75	57	43
Throughput	Department	43	57	100	100
	MAJCOM or NAF	25	50	50	42
	Wing and below	n/a	n/a	45	74
	Non–Air Force	75	75	17	52
Cost	Department	14	57	100	100
	MAJCOM or NAF	50	75	42	58
	Wing and below	n/a	n/a	26	74
	Non–Air Force	0	75	43	52
Objectives	Department	72	57	100	100
	MAJCOM or NAF	100	50	92	42
	Wing and below	n/a	n/a	92	80
	Non–Air Force	75	75	72	29

Understanding Program Objectives and Impacts Is Essential for Assessing Processes and Outcomes

There is a general lack of awareness regarding program objectives, participant views, the effect on partner capabilities, and activity objectives that will limit the Air Force's ability to assess outcomes, design and theory, and important aspects of program processes and implementation. Perhaps most importantly, OET suffers from this deficiency because its stakeholders are generally unaware of the program's overall objectives, possibly because of its singular lack of formal guidance.

Next we will summarize the results by looking at the responses to groups of questions that relate to potential data collector and assessor roles for outcome assessments with respect to OET (Table 3.2) and MPEP (Table 3.3). The upper part of each table shows the results of four questions related to collecting data, and the lower part shows the results of four questions related to assessing data. The percentages in each cell represent the proportion of respondents who could perform either a data collection function or an assessment function. In the case of Table 3.2, these percentages correspond, respectively, to department-level respondents, MAJCOM and NAF respondents, and joint respondents. The color-coding, while subjective, is designed to highlight a high (green), mid-range (yellow), and low (red) percentage of positive

Table 3.2
Operator Engagement Talks Outcome Assessment Analysis

Question	Percentage of Positive Answers, by Level		
	Department	MAJCOM or NAF	Joint
Data Collector			
Question 9: Participates in specific activities	72	75	75
Question 15: Participant views	43	75	75
Question 19: Effect on partner capability	14	25	50
Question 23: Compliance with legal requirements	29	25	25
Assessor			
Question 26: Prepares reports on specific activities	57	50	75
Question 42: Reports on how well specific activities' objectives are met	29	50	50
Question 46: Knows program objectives	0	0	25
Question 47: Sets objectives for specific activities	14	0	0

Table 3.3
Military Personnel Exchange Program Outcome Assessment Analysis

Question	Percentage of Positive Answers, by Level			
	Department	MAJCOM or NAF	Joint	Wing and Below
Data Collector				
Question 9: Participates in specific activities	100	83	88	72
Question 15: Participant views	100	67	34	29
Question 19: Effect on partner capability	100	17	40	29
Question 23: Compliance with legal requirements	100	50	14	29
Assessor				
Question 26: Prepares reports on specific activities	100	42	74	14
Question 42: Reports on how well specific activities' objectives are met	100	8	30	14
Question 46: Knows program objectives	0	0	0	14
Question 47: Sets objectives for specific activities	0	8	8	0

answers for each question. Table 3.3 uses the same color-coding scheme, but adds an additional level of respondent: wing and below.

For example, in Table 3.2, question 9 asks respondents if they participate in specific security cooperation program activities, which might enable them to collect data on program performance. Seventy-two percent of department-level respondents answered in the affirmative, as did 75 percent of both MAJCOM/NAF respondents and joint respondents. Other data collection questions focus on types of data that might assist in understanding how well a program is meeting its objectives (participant views, effect on partner capability, and compliance with regulatory requirements). In the bottom portion of Table 3.2, four questions attempt to identify respondents whose organizations could serve in an assessment role. Question 26, for example, asks if the respondent prepares reports on specific activities; this implies that the respondent may be making some judgments about the activity's effectiveness. Fifty-seven percent of department-level respondents answered yes to this question, as did 50 percent of MAJCOM/NAF respondents and 75 percent of joint respondents. The remaining three questions in the bottom portion of Table 3.2 ask about respondents' knowledge of program and activity objectives.

Our analysis of the responses summarized in Table 3.2 revealed a number of important shortcomings about the OET program. Among Air Force respondents engaged in OET activities at both the department and MAJCOM/NAF levels, none indicated that they have an awareness of the program's overall objectives. Another shortcoming may be the very small numbers engaged in designing specific OET activities. Only a small percentage (14 percent) of the respondents at the department level claim to help set objectives for specific activities, and no respondents at the MAJCOM/NAF level identified this as part of their contribution to the program. Not surprisingly, with little upfront knowledge of the activities' objectives, less than 30 percent of the department-level respondents said they report on how well a specific activity's objectives were met. In addition, less than 60 percent of these respondents prepare reports on specific activities, less than 45 percent said they collect data on participant views, and less than 15 percent collect data on how the program affects partner capability.

As with OET, MPEP respondents demonstrated a lack of awareness and involvement with respect to program objectives. In fact, none of the respondents indicated they knew the overall objectives of the program, which is surprising given that these objectives are explained in AFI 16-107.[6] In addition, many were not aware of or did not collect information regarding specific activity objectives, participant views, or the program's effect on partner capabilities. Finally, less than one-third of MPEP respondents currently report on how well objectives for specific activities are being met.

Understanding Program Resources and Contributions Is Essential for Need and Cost-Effectiveness Assessments

Department-level respondents indicated that they are not well postured to conduct need-for-program or cost-effectiveness assessments. None of the department level respondents said they determine the need for the program, yet half of the MAJCOM and NAF respondents said they make recommendations regarding this aspect of OET.[7]

Our analysis revealed a number of insights about program resources. None of the respondents at department level advocated for funds, and fewer than half of these respondents said they managed resources. The need for a program is inextricably linked to funding, as is funding advocacy, a function normally reserved for program managers at HAF.[8] Although there were a number of respondents collecting resource data for MPEP, very few (less than 20 per-

---

[6]   Paragraph 1.3 of AFI 16-107, *Military Personnel Exchange Program (MPEP)*, February 2, 2006, describes the goals of Air Force MPEP as (1) promote mutual understanding and trust, (2) enhance interoperability through mutual understanding of doctrine, tactics, techniques, and procedures of both air forces, (3) strengthen air force–to–air force ties, and (4) develop long-term professional and personal relationships.

[7]   The survey results discussed in the remainder of this chapter are presented in Table A.2 in Appendix A.

[8]   It is possible that this lack of ownership of the resource advocacy function is a reflection of a less than fully representative sample of departmental responses, especially from senior-level officials.

cent) said they advocate for funds. Moreover, there was little to no insight at any level as to how MPEP compares with other Air Force security cooperation programs. As a result, it is not clear who would conduct an assessment of the program's cost-effectiveness.

Overall, OET stakeholders need better awareness of how the program compares with other, similar, programs, and would almost certainly benefit from a clear advocate. Collectively, addressing these areas could help the Air Force effectively assess the OET at each level of the evaluation hierarchy.

Collecting MAJCOM/NAF and Joint Stakeholder Input Is Essential for All Types of Assessments

The survey responses made it clear that MAJCOM/NAF-level and joint input is essential for all types of assessments. For example, about two-thirds of the MAJCOM/NAF-level respondents engaged in MPEP indicated they collect data on participant views, a key piece of data that can be useful in understanding the effect of a program on partner capabilities. Informed input from those in charge of executing security cooperation activities at the wing level and below is also essential. For example, respondents from the wing level and below indicated that they collect data on how MPEP affects partner capabilities, and 40 percent said they have access to information regarding the effect of MPEP on partner activities. None of these respondents, however, claimed to know the MPEP program objectives—although awareness of program and activity objectives would help these stakeholders gather data that could be used effectively for program assessments.

Similarly, the Air Force could benefit greatly from an effort to actively solicit input from joint stakeholders, which could help to ensure adequate design of MPEP. Forty-three percent of the joint respondents indicated that they are heavily engaged in designing activities, while 57 percent said they manage resources used by the MPEP program. In addition, these joint respondents collect data on requests to participate in MPEP, and these respondents also have access to information on partner capability assessments. However, this is largely a missed opportunity to enhance the Air Force's ability to assess MPEP. Indeed, although more than 70 percent of the joint respondents claim to participate in MPEP activities (and therefore have the potential to collect useful information), only 14 percent said they provide any sort of report on these activities. As one Assessment Day participant suggested, "data is both abundant and hard to get." The sources of data for program managers are not always intuitive, and, as the survey data illustrate, program managers must reach out to a wide variety of stakeholders to get it.

None of the department-level OET respondents indicated that they recommend countries to participate, although 25 percent of the MAJCOM and NAF respondents said they do. Given the relatively low involvement at the department level implied by the respondents, collecting and assessing the data gathered by MAJCOM/NAF and Air Force personnel serving in joint assignments seems important for this program. Our analysis revealed that half of the MAJCOM/NAF OET respondents claim to make recommendations regarding increases and decreases in participation, a key factor in understanding the need for the program, as well as its design and theory. While less than 43 percent of department-level respondents replied that they are involved in designing activities, 75 percent of MAJCOM/NAF and joint respondents indicated that they do contribute to this important activity. Similarly, despite just 43 percent of department-level respondents saying that they see partner requests to participate, and less than 30 percent reporting that they have access to partner capability assessments, 50 percent of MAJCOM/NAF and joint respondents said that they do have access to these types of data.

Aside from the issue of gathering data related to OET activities, MAJCOM/NAF and joint respondents' answers suggest that their respective stakeholder organizations are engaged in key aspects of implementing the program. While less than 45 percent of department-level respondents said that they manage resources, and none advocate for funds, 50 percent of MAJCOM/NAF respondents said they do manage resources, and 25 percent of joint respondents indicated that they advocate for funds that support the program. The result is that program managers do not have full control over the resources that make their programs possible. This issue was also addressed by one of the Assessment Day focus groups, which identified the lack of authority as a "big problem" and noted the obstacles to achieving full cooperation and information access among organizations.[9]

Program Guidance Is Uneven

Program design and implementation guidance can be the "yardstick" for measuring program effectiveness. By asking questions about formal and informal guidance, both within and outside the Air Force, the study team developed insights into where such guidance exists, how it is used, and where the gaps exist.

Air Force Instructions Are Not Always Available and Not Always Used. The issue with Air Force program guidance appears to be largely a matter of training and compliance, in the case of MPEP, and lack of availability, in the case of OET. Of the 70 respondents involved with the MPEP, only 32 indicated that that they were aware of relevant Air Force–issued guidance, in spite of the existence of an AFI relating to the program. Of the 38 respondents who were unaware, most were at the wing level and below (24), with joint respondents largely making up the difference. The situation with OET is different. Only slightly more than 13 percent of the respondents involved with OET indicated that they used Air Force directives in the design and development of that program. However, at the time of the survey, there was no formal AFI governing OET, so the documents referred to by these respondents consisted of memos and policy papers.

Joint/Interagency Guidance Also Applies to Air Force Security Cooperation Programs. Air Force security cooperation officials should be making use of existing joint and interagency guidance—of both a strategic and administrative nature—in the design and development of their programs. However, nearly 85 percent of respondents indicated they did not use such documents. Of those respondents who indicated they do use joint/interagency documents, most were involved with MPEP and cited documents such as bilateral memoranda of understanding, status of forces agreements, and most frequently, Department of Defense Directive 5230.20, *Visits and Assignments of Foreign Nationals*. These documents primarily address regulatory requirements that govern the mechanics of personnel exchanges, although they can also apply to programs like OET that include visits of foreign nationals and U.S. military visits to foreign countries. Appropriately, there were some respondents at all levels who indicated that they collected information regarding MPEP's compliance with legal and regulatory requirements.

Stakeholders Create Their Own Guidance in the Absence of Formal Guidance. Finally, respondents were asked if they used informal guidance documents, such as continuity binders,

---

[9]   Forty-two percent of "Assessment Day" participants indicated there is a need for greater information sharing in order to make assessment possible.

to aid in the design and development of the programs.[10] As might be expected, a substantially larger number of OET respondents indicated such use as compared with those involved in MPEP. Slightly more than 25 percent of MPEP respondents, who do have access to formal Air Force program guidance, indicated they use informal guidance documents, whereas 40 percent of the OET respondents indicated they do. The majority of the respondents using these documents were at the department level, suggesting that it would be possible to formalize such guidance in an official document, such as an AFI.[11]

Assessment Skills Exist but Could Be Improved

To gain insight into the respondents' views of their ability to conduct actual assessments, the RAND study team asked about the skills they possess that could enable them to conduct such assessments. Respondents were asked to comment on existing assessment skills that they, or their successors, might bring to the job, as well as their view of the need for assessment skills training, checklists, and instructions on how to accomplish assessments.

Many Stakeholders Believe They Have the Skills to Conduct Assessments. The first of three questions respondents were asked to comment on focused on the skills they believed they already possessed. Specifically, they were asked, "Do you believe that you, or personnel assigned to your position in the future, have/will have the skills to conduct appropriate security cooperation assessments (e.g., regarding the need for a program, program design, program compliance with policy, program outcomes, and program cost-effectiveness)?" Overall, slightly less than half answered "yes," indicating that many stakeholders believe they have the necessary assessment skills. Interestingly, the percentage was much lower for those respondents involved with the OET program; just one-third believed that they, or their successors, would have such skills. Also, it is unclear whether respondents' views of their assessment qualifications should be taken at face value, given the limited understanding of assessment roles and missions revealed by participants in RAND's Assessment Day focus groups (see Chapter Two) and the lack of official qualifications for those responsible for conducting Air Force security cooperation assessments.

An Assessment Instruction and Checklist Could Improve a Stakeholder's Ability to Assess. One view expressed during Assessment Day was that if the Air Force wants to assess security cooperation programs, it should develop guidance and instructions for their conduct.[12] To understand what it might take to prepare a potential assessor, the survey respondents were asked, "If no, do you believe that you, or personnel assigned to your position in the future, would be prepared to conduct assessments if you had an appropriate checklist and set of instructions?" About half of those who had answered "no" to the first question answered "yes" to this question, suggesting that developing an assessment instruction and accompanying assessment checklists might be an effective way to prepare those involved with Air Force security cooperation programs to conduct assessments of them. The similarity of such an approach

---

[10] We acknowledge that the distinction between formal and informal guidance is not always entirely clear. For example, continuity binders often contain extracts from official guidance documents.

[11] At the time of this writing, HAF Regional Affairs Division, which manages the OET program, was preparing such an instruction for the program.

[12] Fifty-seven percent of the "Assessment Day" participants commented on the need for guidance in order to perform appropriate assessments. This was identified as the single largest impediment to conducting sound assessments.

with the current Air Force Inspector General approach mentioned earlier is strong; it is an approach that airmen are familiar with and have confidence in.

Conclusions

In general, the analysis suggests that there are stakeholders throughout the Air Force, as well as Air Force personnel serving in joint assignments, who have the potential to both collect data and assess for Air Force security cooperation programs. Moreover, more than half believe they have the skills to conduct these assessments, and another quarter believe they can do it with the right instructions and checklists.

However, this analysis is not intended to understate the impediments to implementing an effective assessment program. Clear and consistent program guidance needs to be developed in some cases, and disseminated in all cases. In particular, potential data collectors and assessors need to be aware of overall program and individual activity objectives. Stakeholder coordination and information sharing could also be improved. Reporting requirements, if they exist, may not be fully understood or complied with. Furthermore, clear program advocates with the right authorities are essential for implementing the assessment framework. Finally, the findings in this chapter are largely based on survey information collected from stakeholders associated with two Air Force–managed programs, supplemented by observations provided by Air Force officials during RAND's Assessment Day. A larger survey of participants in other programs managed and/or executed by the Air Force will be needed to gain a detailed picture of the service's current and potential capacity to conduct a full range of security cooperation assessments.

Assuming that sufficient assessment capacity exists within the Air Force security cooperation community, how should the Air Force move toward a consensus on implementing a comprehensive and integrated assessment system? That is the subject of Chapter Four.

Strengthening the Case for a Comprehensive Approach to Security Cooperation Assessment

The previous chapters have examined attitudes within the Air Force's diverse security cooperation community regarding the advisability of a comprehensive approach to assessment. It is clear that considerable doubt exists among many stakeholders that such an approach would be beneficial, or even doable, in spite of data indicating that the Air Force probably has the capability to conduct a wide range of assessments with proper guidance and management. This chapter examines the case for enhanced security cooperation assessments, with the intent of establishing the basis for a stand-alone assessment instruction or annex to the Air Force Campaign Support Plan or the Air Force Global Partnership Strategy. Such high-level guidance, we believe, would help convince doubters regarding the need for comprehensive assessment and the ability of the Air Force to undertake such an initiative without a significant increase in resources.

Drawing on the RAND assessment framework developed in 2008, this chapter suggests answers to four elemental questions about assessment:

- Why assess?
- What to assess?
- How to assess?
- Who should assess?[1]

By addressing these questions, this chapter will show that the RAND security cooperation assessment framework can help substantiate the assumptions that Air Force security cooperation efforts can benefit if there is a common understanding of security cooperation assessments; and that security cooperation assessments can support DoD and Air Force decisionmaking if they are available, current, accurate, and configured appropriately. To do so, this chapter begins with a broad, general examination of each of the elemental assessment questions listed above. It then applies each assessment element to two specific security cooperation examples, MPEP and OET,[2] which were the main focus of Chapter Three. This chapter concludes with a discussion that suggests how enhanced security cooperation assessment might be more deeply embedded into Air Force management processes. By the end of the chapter, readers should have a clear sense of the potential benefits the Air Force could realize from enhanced security cooperation assessments.

---

[1] Moroney et al., 2010a.

[2] OET was previously called the Operator-to-Operator Program or Operator-to-Operator Staff Talks.

Why Assess?

The answer to the question "Why assess?" is rooted in the Air Force's desire to improve partnership designs in order to enhance their effectiveness for the United States and to its partners; to reveal opportunities to improve Air Force force development and force employment; and ultimately, to manifest a beneficial effect on overall force management decisions.

Although the Air Force already undertakes security cooperation assessments, especially partner needs assessments and program performance assessments,[3] the Air Force has five broad motives to expand its assessments of its security cooperation efforts. These are a desire to

- improve partnership design
- support accurate partner prioritization
- contribute to force development decisions
- contribute to force employment decisions
- assist force management decisions.

Each of these motives will be examined in the following sections.

Improving Partnership Design

The Air Force's efforts to improve the design of its partnerships with allied or friendly air forces are driven by the goal of helping the United States and its partners expeditiously progress toward the strategic end states found in the Air Force Global Partnership Strategy.[4] Accordingly, some security cooperation events, programs, and activities will be more beneficial than others for a given country, at a given point in time and for a given purpose. For example, a high-profile visit by senior Air Force leader may be appropriate for helping to establish a relationship with a new or politically sensitive partner, whereas a bilateral or multilateral exercise may contribute more to interoperability objectives with long-standing coalition partners, and providing aircraft maintenance training may be appropriate for partners whose aviation capacity the United States wants to improve. Therefore, efforts to fine-tune a partnership's design should be an important consideration for the Air Force, since a more effective design should ultimately lead to the more efficient use of Air Force security cooperation resources.

Partner Prioritization

In prioritizing partnership-building efforts, U.S. officials should consider both the partner's strategic value to the United States and its ability to exploit U.S. assistance. Some partners are more important to the United States than others, given their geographical position in the world, the confluence of their interests with those of the United States, and their relative military prowess. Among the partners with whom relations and military interoperability are less evolved, there is also the question of relative progress: Which partner is making the greatest strides toward the strategic end states in the Air Force Global Partnership Strategy?

---

[3]   Based on discussions and interviews with USAF security cooperation stakeholders in HAF and the Office of the Secretary of the Air Force, MAJCOMs/NAFs, regional components, and wing- and lower-level units—as well as security cooperation officials in other military departments, COCOMs, OSD, the Joint Staff, and U.S. embassy country teams.

[4]   The Global Partnership Strategy may be understood as a point of guidance consolidation, because it captures the essence of national and DoD-level guidance (e.g., Guidance for Employment of the Force and Guidance for Development of the Force) and leavens this higher-order guidance with Air Force–specific instructions of its own. See U.S. Air Force, 2008.

The strategy contains four such end states, three of which apply to partners and a fourth that has to do with Air Force force development capabilities:

1. Establish, sustain, and expand global partnerships that are mutually beneficial.
2. Provide global partners the capability and capacity necessary to provide for their own national security.
3. Establish the capacity to train, advise, and assist foreign air forces while conducting partnership activities, using USAF airmen with the appropriate language and cultural skills.[5]
4. Develop and enhance partnership capabilities to ensure interoperability, integration, and interdependence, as appropriate.

An enhanced security cooperation assessment capability will help the Air Force to prioritize its partners properly. Knowing which partners are making the greatest progress toward given strategic end states will offer the Air Force opportunities to maximize efficiency in the distribution of security cooperation resources. It also will support efforts to identify the highest-performing partners in areas of high interest to the Air Force and better enable the Air Force to reinforce success by apportioning more resources to security cooperation activities with those high-performing partners.

Of course, the Air Force will need to maintain a balance between rewarding superior performance and absorptive capacity and supporting partners with relatively greater needs and strategic value to the United States. The latter could be defined in terms of force employment (their potential to contribute to air, space, and cyberspace resources to achieve U.S. national security, DoD, and COCOM objectives), force development (their potential to contribute to the furtherance of U.S. Air Force capability requirements, for example, through joint acquisition or cooperative research and development activities), and force posture (their potential to provide bases and overflight rights to support U.S. air forces operating overseas).

Force Development

In the security cooperation realm, the following strategic end state drives the force development process: capacity to train, advise, and assist foreign air forces, while conducting partnership activities using USAF airmen with the appropriate language and cultural skills. Assessments can support Air Force force development by helping the service to calibrate security cooperation supply and demand. If the Air Force was able to reliably anticipate the demand for security cooperation (i.e., qualified airmen, specific exercises, activities, events, and programs) from various partner countries, then AETC and the Deputy Chief of Staff for Manpower and Personnel (AF/A1) could plan accordingly and alert the Chief of Staff of the Air Force and the Secretary of the Air Force to the need for more money, manpower, and training to generate the appropriate security cooperation capacity. The accuracy of annual estimates for next year's security cooperation capacity could be improved through sound supply and demand data that aggregate COCOM requirements and AETC's capacity to produce security cooperation capabilities of the type sought by partners. For example, although it is clear that the COCOMs

---

[5]   Although this end state is largely inward-focused in nature, i.e., declaring that the Air Force will prepare its personnel (a force development goal, as opposed to a security cooperation goal), it does state that it will conduct security cooperation activities using these personnel as key resources.

would like more language-qualified airmen, it is not clear how many they need or what languages they should speak. Nor is it apparent how much DoD's educational capacity should expand in order to satisfy COCOM demand for more linguistically proficient personnel.

Force Employment

The Air Force should pursue assessments to determine whether the employment of Air Force resources in building partnerships is meeting its expectations for helping partners reach the service's strategic end states,[6] as well as the objectives spelled out in the Guidance for Employment of the Force and the COCOMs' Theater Campaign Plans. Some security cooperation activities and programs are more effective at this than others, depending on the partners involved and the circumstances. Assessments could identify the high-performing security cooperation activities and programs that are delivering the expected progress of the various partners. Once the Air Force knows which efforts typically produce the most results, it could reinforce those more successful security cooperation programs in its force employment decisions. Thus, for example, the service might find that a particular exercise series is most effective in producing improved partner air sovereignty capabilities. Therefore, the Air Force might allocate more resources to that exercise series than to another security cooperation activity that usually yields less-impressive results. Of course, the appropriateness of a particular security cooperation activity will depend on the characteristics of the partner in question, such as its level of capacity, its willingness to engage with the United States, and its strategic importance to the United States. In addition, the Air Force needs to consider that certain security cooperation activities tend to build on one another—for example, successful pilot or aircraft maintenance training in the United States is contingent on the foreign partner's proficiency in English.

Force Management Processes

Here, the Air Force should seek to evaluate the degree of coherence between its security cooperation assessments and DoD management processes, such as the PPBE system and the Joint Capability Integration and Development System. It is our assumption that, if the level of coherence is high, management processes will benefit. Given this reasoning, SAF/IA and HAF, especially AF/A3/5 and the Deputy Chief of Staff for Strategic Plans and Programs (AF/A8), should want to know whether security cooperation assessments are available, current, and accurate in order to support many of their management activities, including strategic planning (e.g., Where do we need additional partners?) and budgetary and programmatic decisions (e.g., What resources do we apportion to U.S. Air Forces in Europe for security cooperation in the program objective memorandum [POM]?). In this regard, the Air Force should also have security cooperation assessments available at the appropriate times to support deliberations within the Air Force Corporate Structure as well as between the Air Force leadership and elements of the joint community with an interest in security cooperation and security assistance. Currently, these deliberations are being conducted without a comprehensive and detailed picture of the amount resources the Air Force is devoting to security cooperation or the COCOMs' requirements for Air Force security cooperation activities.

---

[6]   These are the strategic end states in the Air Force Global Partnership Strategy, as listed earlier in this chapter.

What to Assess?

The key question "What to assess?" attempts to identify observations and information that provide insights about security cooperation partnership design and performance, security cooperation force development and employment, and force management decisions. Although there is a broad menu of potential answers to this question, the Air Force should generally look for observations and information that can, either directly or indirectly (i.e., by inference or deduction), facilitate measurement of security cooperation–related evidence to inform the five categories of decisionmaking noted in the "Why assess?" section above. In this regard, those involved in Air Force security cooperation management and execution should systematically collect resource and participant data on security cooperation events, programs, activities, and initiatives, as well as U.S. and foreign participants' reactions and valuations of them and specific partner country actions that can reasonably be attributed to them.

The Air Force organizations collecting and assessing the data, the decisionmaking process they are supporting, and the requesting official's or organization's requirements will determine the activities assessed, the data collected, and the form of analysis selected. For example, Air Force units involved in the execution of ongoing activities with specific partners may be primarily interested in demonstrating the near-term results of their activities by country to the appropriate regional component or NAF. For their part, regional components may have an interest in aggregating data related to the security cooperation activities they manage according to the Campaign Plan objectives for which they are responsible. This process of data collection could support COCOM assessments of regional and country security cooperation performance over a period of time, as well as HAF and MAJCOM requirements for information on security cooperation resource expenditures and shortfalls. Finally, SAF/IA program managers and regional desk officers have their own unique data needs. In the future, the former will have to show how their programs support the goals outlined in the Air Force Global Partnership Strategy and the Air Force Campaign Support Plan—as well as to continue to justify their budgetary requests within the Air Force Corporate Structure. And regional desk officers will continue to face requests from Air Force leadership for a variety of political-military information on foreign country leaders, developments, and activities. In each case, security cooperation data will need to be collected, combined, and assessed in different ways.

How to Assess?

To address the question of "How to assess?" the Air Force should understand four elements that will dictate the answer: assessment types, assessment objectives, assessment effects, and assessment data and observations.

Assessment Types

Assessments may feature a variety of attributes. First, an assessment could be quantitatively oriented and based on things that, when counted, yield some useful result. In contrast, an assessment could be qualitatively oriented, reflecting characteristics and attributes that do not lend themselves to quantitative methods or that are simply more revealing in their own right. In either instance, assessments also could be either direct or indirect (measuring proxies or intermediate variables). Finally, assessments may also be single-point, "snapshot in time" evaluations that reflect the status of a particular security cooperation activity or event, or they could be a time-series assessment that aggregates multiple earlier assessments to reveal a trend. To a large extent, the type of assessment will depend on the objective, as defined in the abstract by RAND's hierarchy of evaluation.

Assessment Objectives

Assessments may have different objectives, spanning the hierarchy of assessments that emerged from the RAND assessment framework study.[7] Some assessments may seek to determine or validate the need for a specific security cooperation program, event, or activity. Others may examine the security cooperation program's design and theory to see whether it is reasonable to expect the program to develop the partner or to produce the outputs or outcome as intended. Other assessments may focus on determining the security cooperation activity or program performance in terms of outcome or output. Still others may examine the process and implementation in order to determine the degree to which it complies with its governing instructions, directions, and regulations. Finally, there are cost and cost-effectiveness assessments. In the former, the objective is to ascertain the costs of a given security cooperation program, event, or activity, while the latter performs a comparison among programs, events, or activities, thereby determining their relative cost-effectiveness.

Assessment Effects

Assessments can have different effects as well. As the earlier RAND study pointed out, assessments are meant to inform decisions. In general, these decisions relate to choosing and prioritizing partners; developing appropriate security forces to build foreign partnerships; employing security cooperation forces within the context of particular activities, programs, and initiatives; and managing security cooperation forces through DoD's PPBE and force allocation/provision processes. Thus, assessment effects are intended to provide information that supports these decisionmaking and managerial processes. For example, partner selection and prioritization decisions should ideally rely on the results of assessments that integrate and standardize information on the military capabilities and capacity of potential partners, as well as their strategic importance to the United States and their willingness to engage with the United States, among other things.

Assessment Data and Observations

Data are the fuel on which assessments run, and they can take many forms. They might be simple observations, such as watching mechanics maintain an aircraft engine and reaching conclusions about their skills. Or they might be impressions that result from close working relationships formed between U.S. and partner airmen during an exercise, which enable the U.S. airmen to form some sense of the partner's overall proficiency in their jobs. Data might be simple numerical figures, such as the number of partner students that were enrolled and the number that graduated. Data can take the form of cost figures and cost per episode, per event, per item, per capita, etc. Or data can be the often-dismissed anecdote, which, in reality, may be a field observation and a valid data point if collected by an airman with the appropriate expertise. To the extent possible, data should be collected, reported, analyzed, validated, and

---

[7]   Moroney et al., 2010a.

integrated in a standardized manner in order to support the accuracy, reliability, and comparability of program and country assessments over time.

Who Should Assess?

As a general proposition, assessments should be performed by the personnel who are closest to the event, activity, program, or country under scrutiny, as long as they also possess the necessary skills, experience, and objectivity to conduct the assessment. That said, at least three considerations have bearing on the answer to the question of "Who should assess?"

First, there are the various assessment functions or roles to be filled. Qualified personnel are needed to gather data, conduct the actual assessment, check and validate the assessment, integrate the results with other assessment results, and, finally, present them to decisionmakers. The RAND assessment framework study labels these roles as data collector, assessor, validator, integrator, and recommender.[8]

- Primary data collectors should be directly involved in executing, participating in, or observing security cooperation events. Although they do not normally require formal assessment training, they should be familiar with the objectives of the event, as well as with program and/or regional/country goals that pertain to the event. They should also possess a clear set of guidelines (e.g., a checklist) that defines the kind of data they are tasked with gathering. Primary data collectors may report to secondary data collectors—often program or component officials who are responsible for aggregating information from numerous security cooperation events. These officials or others within the same organization may also play an assessment role.
- Ideally, assessing organizations should be staffed with security cooperation planners and programmers, country experts, and operations researchers or others with formal training in quantitative and qualitative evaluation methods.
- Validators, almost always from a different organization from the one responsible for assessment, do not need to have detailed program or country expertise. However, they should have a clear understanding of Air Force and other relevant security cooperation objectives and measures that relate to the program or country under evaluation, as well as the ability to judge the quality of the analysis presented by the assessment organization.
- The integrating organization should have information technology and knowledge management experts capable of storing, compiling, and sorting a considerable quantity of program- and country-based assessment information. It should also have strategic planners who can synthesize this information in support of high-level Air Force decisions and recommendations.

Second, in many cases, non–Air Force organizations may provide the best source of assessments. The U.S. Army, Navy, Coast Guard, and Marine Corps may also be involved. Since partners also have relationships with the Office of Defense Cooperation at the U.S. embassy, Security Assistance Office and attaché personnel can also watch events, participate in exercises, and otherwise become involved with security cooperation assessment activities. The COCOM

---

[8]   Moroney et al., 2010a.

and air component staffs may also be involved. Thus, a security cooperation assessment may become a joint affair, and Air Force security cooperation assessments may require the cooperation of Air Force personnel serving in joint assignments to collect data, perform some aspect of assessment, or support assessment validation.

Third, because the Air Staff and SAF/IA do not directly supervise many security cooperation activities, they should clearly define supporting-supported relationships between HAF, functional MAJCOMs, regional components, NAFs, wings and below—as well as joint organizations, such as U.S. embassy security assistance offices and defense attaché organizations and COCOM staff offices—to collect the data and make observations, gather statistics, or conduct the interviews necessary to perform a given assessment. These supporting-supported relationships will often be driven more by proximity and access to the partner and security cooperation events than by the otherwise established chain of command. These supporting-supported relationships, set up to facilitate security cooperation assessment, may need to be codified in memoranda of understanding, or some similar agreement, to ensure a stable relationship that produces consistent security cooperation assessments over time.

Implementing Country-Oriented Security Cooperation Assessments

In this section, we identify the key players and their relationships to the measures or criteria for which data must be collected, and the general classes of assessments that would be involved. In our view, this construct could be applied to security cooperation efforts that are either country-oriented or program-oriented—although the specific unit of analysis would depend on the security cooperation function being examined, as well as the assessment focus and organization performing the assessment.

This construct is illustrated in Table 4.1, which indicates that there are four security cooperation functions that can benefit from assessment: partnership design and prioritization, force development, force employment, and force management. The focus of assessments that generally support these functions appear under the "Assessment Focus" column, and the "Type of Assessment" column reflects the level in the hierarchy of assessments. The "Assessment Roles" column suggests what entity might serve in the stakeholder roles that were identified in our previous assessment study. The "Decision Category Informed" column suggests the nature of decisions (strategic, programmatic, or good order and discipline) that the assessments might inform.

A key insight from Table 4.1 is that, for country-oriented force employment assessment and part of the country-oriented force development assessment, the Air Force need not evaluate each program in order to generate useful assessments. Broader metrics will often suffice to inform these evaluations.

Partnership Design and Prioritization Assessments

As Table 4.1 suggests, the focus of partnership design and prioritization is on the partner, specifically its suitability for engaging in security cooperation activities with the United States. A partner's security cooperation suitability can be understood in terms of its air force's capabilities and the country's acceptability in terms of its human rights, rule of law, status as a democracy, and policy alignment with U.S. interests.

Table 4.1
Security Cooperation Assessment Construct

Security Cooperation Function	Assessment Focus	Type of Assessment (from the hierarchy of evaluation)	Assessment Roles	Decision Category Informed
Partnership design and prioritization	Partner capacity, acceptability, alignment with U.S. interests	Threat, needs, capability, political-military suitability	Data collector, assessor = air component Validator = AF/A3/5 Integrator = SAF/IA	Strategic: Is partner selection consistent with U.S. interests?
Force development	DOTMLPF adequacy regarding demand for security cooperation capability, security cooperation programs	Resource, design, process compliance	Data collector, assessor = COCOM, air component, AETC Validator = AF/A1 Integrator = SAF/IA	Programmatic: Is investment in force development sufficient to satisfy demand for security cooperation? Good order and discipline: Were all rules obeyed?
Force employment	Programs, partners, security cooperation activities	Outcome/impact, process compliance	Data collector, assessor = COCOM, air component Validator = AF/A3/5 Integrator = SAF/IA	Strategic: Are partners and programs taking the U.S. closer to desired end states? Good order and discipline: Were all rules obeyed?
Force management	PPBE, force provision/ allocation	Cost, cost-effectiveness, risk	Data collector, assessor = air component Validator = COCOM Integrator = SAF/IA	Programmatic: Are security cooperation investments paying off, are outcomes cost-effective, do the investment choices limit risk?

U.S. decisions concerning partnership design and prioritization are or should be informed by assessments of the potential partner's defense needs—the threats to the state's security and the resulting national security needs, or the partner state's current defense capabilities relative to the threat—and the partner's political-military suitability. Political-military suitability might be assessed by evaluating the partner's observation of norms in international relations: respect for treaties and state boundaries, observance of international law, and a healthy civil-military relationship at home, including a favorable human rights and rule of law record.

Consistent with the belief that assessments should be performed as far forward as possible, the air component command might serve both as data collector and as assessor. The command's intelligence directorate, availing itself of Defense Attaché Office and Office of Defense Cooperation information and other in-country reporting (e.g., 6th Special Operations Squadron end-of-mission reports) would be one candidate for collector and assessor. AF/A3/5 might serve as the assessment validator, engaging the Office of the Deputy Chief of Staff for Intelligence, Surveillance and Reconnaissance as appropriate. SAF/IA would perform the assessment integration role, probably integrating judgments from the air component assessment with those from the State Department, U.S. Agency for International Development, and other executive branch views of the potential partner's worthiness for closer engagement through security cooperation activities. The result would be to inform U.S. strategic decisionmaking and to determine whether selecting this state as a partner is consistent with U.S. interests and whether there is a reasonable expectation that the partner can become a security contributor (rather than a consumer) in a reasonable amount of time.

Force Development Assessments

Force development assessments would use supply and demand for properly qualified airmen (Global Partnership Strategy strategic end state 3) as the primary, although not the sole, criteria for assessment. This measure would be complemented by assessments of the adequacy of the DOTMLPF for generating and sustaining the supply of properly qualified airmen. On the demand side, the COCOMs and their air components probably would have the best access to information regarding the need for security cooperation activities in the country, so they would serve as data collectors for demand. The Air Force's AETC and AF/A1 would have the best access to information on supply, so they would be the logical data collectors for that.

AETC and AF/A3/5 would appear to be the best-suited organizations for conducting force development assessments. They would be aware of the supply of building partnerships capabilities and would have some sense of how security cooperation fares in the Air Force's overall priorities. However, not all force development assessments would fall to these organizations. Indeed, a comprehensive assessment of the Air Force's capability and capacity to support security cooperation would involve much of the Air Staff to assess DOTMLPF by answering questions such as the following:

- Does doctrine support presentation of forces appropriate and sufficient for demand, including the right capabilities and skills, in appropriate numbers, in accordance with the Air Force Building Partnerships Core Function Master Plan?
- Are the resulting forces organized appropriately for the enemy, weather, terrain, partner, and tasks they must perform?
- Does training and education prepare them for success under these conditions?
- Are they equipped appropriately for the enemy, weather, terrain, partner, and prevailing social-cultural context in which they will operate?
- Are their leaders appropriately prepared?
- Are qualified personnel present in sufficient numbers relative to the demand for them?
- Does the Air Force possess sufficient facilities (schools, ranges, laboratories, airfields, etc.) to support security cooperation efforts?

Many of these questions may be answered through end-of-tour questionnaires or reports. The airmen who have just served in these various capacities are best positioned to offer insights and recommendations on the sufficiency of the force development process to prepare them, their equipment, and their subordinates for these assignments.

For the same reasons noted above, SAF/IA would serve as the integrator for force development assessments. It would know where the information contained in the assessment can be used to help the Air Force make the best possible strategic, programmatic, and administrative decisions.

Force Employment Assessments

The partner's progress is measured in terms of force employment through the Global Partnership Strategy's strategic end states. The actual assessment process was summarized above in the discussion of how to assess. The data collectors would be personnel, perhaps assigned to the

COCOM staff but probably in the COCOM's air component command, who are in a position to observe and collect appropriate information. The assessors for air force–related security cooperation activities should also likely come from the air component command. They would be selected for their expertise, which would enable them to evaluate the partner's progress toward

- establishing, sustaining, and expanding relationships that are of mutual benefit to the partner and the United States
- promoting appropriate civil-military relations
- coordinating with key allies and/or partners to assist other states
- growing the capacity for self-defense in terms of air, space, and cyberspace capabilities
- developing and enhancing capabilities to ensure interoperability, integration, and interdependence.[9]

The validators would probably come from the AF/A3/5 staff, or perhaps from SAF/IA. Either office might have the expertise and broad perspective that would enable them to validate the assessment.

SAF/IA would also serve as the assessment integrator. Assessment integration would be organizationally consistent with its other security cooperation–related responsibilities. Moreover, SAF/IA has the insights into Air Force management and programmatic practices that would enable it to do the best job of forwarding security cooperation assessments as inputs for appropriate deliberations and decisions.

Force Management Assessments

Force management assessments generally support efforts to ensure security cooperation planning, programming, and budgeting decisions are sound, meaning that the security cooperation programs are cost-effective, minimize risk, and achieve partner outcomes or program outputs. Force management attempts to align resources with programs, limiting losses where programs underperform, and exploiting success and opportunities among those that perform well.

As Table 4.1 indicates, force management assessments concentrate on the cost and cost-effectiveness level in the hierarchy of evaluation. These "bang for the buck" assessments allow managers to know how their programs are performing and can inform future budgeting and programming decisions by Air Force senior leaders, especially where to look for economies, where to cut, and where to reinforce with additional dollars.

The data collectors and assessors would probably be part of the air component command. This observation may seem counterintuitive because the program offices appear to be the logical choices. However, security cooperation force management tends to be country-focused: The COCOM tends to manage security cooperation on the basis of the countries within the area of responsibility, and SAF/IA also tends to be country-oriented. The Air Force should, therefore, view a partner holistically and consider its overall progress toward the strategic end states, which serve as the principal measures of partner performance.

If the air component command serves as the data collector and assessor, the COCOM is probably the optimum validator because the COCOM staff can pull assessments of the partner from all defense domains (air, land, sea) and assess the partner's relative security cooperation

---

9 U.S. Air Force, 2008.

standing within the area of responsibility. SAF/IA would be the logical integrator, as in the other security cooperation functions described earlier in this chapter.

Force management assessments would inform programmatic decisionmaking. The assessments would help senior leaders identify high-payoff partners (those for which security cooperation investments have produced significant results) and high-payoff programs (those that appear to produce sound results in a variety of different partner countries). In aggregate, such assessments would help identify cost-effective outcomes and offer insights into the partners and programs that should be pursued, as well as those partners that might receive smaller future investments. Finally, such assessments could contribute to a better understanding on the part of the Air Force leadership of the efficacy of security cooperation efforts relative to other ways to achieve U.S. objectives: for example, building partner capacity versus providing U.S. capabilities to establish regional security and stability.

Program-Oriented Assessments

Program-oriented assessments differ from country-oriented assessments in that they collect cost, performance, and outcome data from multiple countries to arrive at judgments about the program in terms of the entire hierarchy of evaluation: whether or not there is a need for it; whether or not its design is sound and likely to yield the desired results; whether or not it is performing (making progress); whether or not its processes operate in accordance with its directives, guidance, and limitations; and whether or not it is cost-effective. Of course, such assessments should take contextual factors into account. Although a program may perform well in one country or historical period, it may not perform well in another place and time.

Program-oriented assessments can occur all along the hierarchy of evaluation. Measures of a program's need might include demand for its activities among countries within the COCOMs. Evidence of the soundness of its design and theory might be found in the performance of the participating partners: If most experience positive results (i.e., progress toward the Global Partnership Strategy strategic end states or toward specified program objectives), the design is probably sound. Process and implementation continues to be gauged against compliance with regulations and instructions. Cost-effectiveness is probably measured in terms of return on investment within the program.

The programs themselves are the logical choice for data collection. They maintain the records and the contacts within COCOMs that would most likely produce the requisite information to fuel the assessment.

AF/A8 and AF/A9 may, in some cases, be the best choices to perform and validate the assessments, as both offices have significant experience and expertise in this regard. SAF/IA would integrate the results with other security cooperation assessments and package the results as appropriate for Air Force and DoD decisionmakers.

Examples of Program Assessment Roles

Tables 4.2 and 4.3 illustrate potential roles for assessing the two AF-managed programs discussed in Chapter Three—MPEP and OET—using the construct presented in Table 4.1 of the current chapter and survey data from Chapter Three. To interject as much objectivity into the

Table 4.2
Proposed Assessment Roles for MPEP

Building Partnerships Function	Assessment Focus	Type of Assessment	Assessment Roles
Partnership design and prioritization	Partner capacity, acceptability, alignment with U.S. interests, threat, needs, capability, political-military suitability	Need for program, design/theory	Data collector = SAF/IAP, AF components, RPMOs Assessor = SAF/IAP, RPMOs Validator = SAF/IAG Integrator = SAF/IAG
Building partnerships force development	DOTMLPF adequacy regarding demand for building partnerships capability, building partnerships programs	Design/theory, process/implementation	Data collector = SAF/IAP, AF components, RPMOs Assessor = SAF/IAP, AF components Validator = SAF/IAG Integrator = SAF/IAG
Building partnerships force employment	Programs, partners, building partnerships activities	Design/theory, process/implementation, outcome/impact	Data collector = SAF/IAP, AF components, RPMOs Assessor = SAF/IAP, AF components, RPMO Validator = SAF/IAG Integrator = SAF/IAG
Building partnerships force management	PPBE, force provision/allocation	Outcome/impact, cost-effectiveness	Data collector = SAF/IAP, AF components, RPMOs Assessor = SAF/IAP, AF components Validator = SAF/IAG Integrator = SAF/IAG

NOTE: RPMO = regional project management office.

process as possible, we find that it is important to have different stakeholders responsible for several of these roles—especially with regard to the assessing and the validating stakeholders. Even if the two roles are most appropriately performed by the same organization, at a minimum, two different offices should be involved in the process. Thus, in the case of MPEP, we recommend that the Policy Directorate within SAF/IA (SAF/IAP) take on the primary assessment role, while the Strategy and Long Range Planning Directorate (SAF/IAG) assumes the responsibility for validation and integration of program assessments. In the case of OET, we recommend that SAF/IA be given a role in validating and integrating assessments conducted jointly by the Air Force component commands and the Plans and Requirements Division of AF/A5XX.

As discussed in Chapter Three, the survey data identified several key points that pertain to assessment stakeholder roles. First, potential data collectors and assessors are largely in place for both MPEP and OET. These include SAF/IAP and the RPMOs for MPEP, and AF/A5XX for OET. However, department-level respondents indicated that they are not well postured to conduct need-for-program or cost-effectiveness assessments. Therefore, the regional air components are probably the most appropriate stakeholder to take the lead in performing these types of assessments.

In addition, the survey responses made it clear that MAJCOM/NAF-level and joint input is essential for most assessments. In Tables 4.2 and 4.3, no joint stakeholders are identified

Table 4.3
Proposed Assessment Roles for OET

Building Partnerships Function	Assessment Focus	Type of Assessment	Assessment Roles
Partnership design and prioritization	Partner capacity, acceptability, alignment with U.S. interests, threat, needs, capability, political-military suitability	Need for program, design/theory	Data collector = HQ AF/A5XX, AF components Assessor = HQ AF/A5XX, AF components Validator = SAF/IA, HQ HAF A3/5 Integrator = SAF/IAG
Building partnerships force development	DOTMLPF adequacy regarding demand for building partnerships capability, building partnerships programs	Design/theory, process/implementation	Data collector = HQ AF/A5XX, AF components Assessor = HQ AF/A5XX, AF components Validator = SAF/IA, HQ HAF A3/5 Integrator = SAF/IAG
Building partnerships force employment	Programs, partners, building partnerships activities	Design/theory, process/implementation, outcome/impact	Data collector = HQ AF/A5XX, AF components Assessor = HQ AF/A5XX, AF components Validator = SAF/IA, HQ HAF A3/5 Integrator = SAF/IAG
Building partnerships force management	PPBE, force provision/allocation	Outcome/impact, cost-effectiveness	Data collector = HQ AF/A5XX, AF components Assessor = HQ AF/A5XX, AF components Validator = SAF/IA Integrator = SAF/IAG

since MPEP and OET are solely Air Force–managed programs. However, COCOM, Joint Staff, OSD, and in-country team feedback should definitely be sought and taken into account. The Air Force could benefit from an effort to actively solicit input from joint stakeholders, which could help to ensure adequate design of MPEP, in particular.

Finally, the lack of awareness regarding guidance, program objectives, partner country views, the effect on partner capabilities, and specific activity objectives is likely to limit the Air Force's ability to assess outcomes, design and theory, and aspects of program processes and implementation. As the survey data reveal, OET stakeholders suffer the most because they are largely unaware of the program's overall objectives, possibly because of its singular lack of formal guidance. However, we note that in February 2010, an AFI on OET was approved that includes data on specific program objectives, resources, and partner country participants.[10]

---

[10] See recommendations made in Moroney, Jennifer D. P., Nora Bensahel, Dalia Dassa-Kaye, Heather Peterson, Aidan Kirby Winn, and Michael J. Neumann, *Enhancing the Effectiveness of the U.S. Air Force Operator-to-Operator Talks Program*, Santa Monica, Calif.: RAND Corporation, TR-805-AF, 2010b, not releasable to the general public..

Enabling Informed Security Cooperation Decisions

Building on the above examples, this section explains how assessments, once completed, might be employed to inform Air Force security cooperation decisionmaking. If security cooperation assessments are properly designed and synchronized with Air Force strategic, programmatic, and administrative processes, they should benefit the Air Force and support its decisionmaking across all levels of the hierarchy of evaluation. Where budget and POM processes are concerned, Air Force decisionmakers will need security cooperation assessment results to help the service determine with which partners to grow, with which partners to maintain the status quo, and with which partners to cut back. The criteria established in the Global Partnership Strategy provide the basis for these decisions as country-oriented security cooperation assessments allow the service to gauge the relative progress in civil-military relations, popular legitimacy, and relative progress in cooperation to assist third parties.

The Air Force will also need risk assessments for those partners that are unable to provide fully for their own national security (also a criterion from the Global Partnership Strategy). Such risk assessments will help the Air Force to identify enduring shortfalls, potential U.S. Air Force gap-fillers, and opportunity costs and regret attending the gap-fillers. The regional air components are probably best positioned to provide these country risk assessments with input from members of the embassy country teams. These assessments could be vetted by SAF/IA and AF/A3/5 and incorporated into the country pages of the Air Force Campaign Support Plan.

On the force development side, the Air Force will want to know the difference between the supply and demand for airmen trained for security cooperation, specifically COCOM requirements measured against Air Force resources.

In a similar vein, the Air Force will want to know which partners prove relatively more (and relatively less) interoperability with it. This answer will also reside—at least in part—in security cooperation assessments in the form of exercise evaluations and impressions from training and assistance events for partners that do not participate in OET.

Program-centric assessments will do what country-oriented assessments cannot: help the Air Force understand which security cooperation programs are providing the greatest value, as well as help the Air Force decide how to manage its programs in the future, both in terms of their funding and their scope. Programs that the senior leadership believes are delivering the greatest contribution to U.S. airpower, based in part on the program-centric assessments described above, would become centers of management attention and priorities for funding. Programs found to be underperforming might become bill-payers.

An important aspect of program-centric assessments is that they support comparisons across programs, enabling officials to see clearly what capabilities each program provides, whether it delivers security cooperation–qualified airmen, beyond-line-of-sight missiles, F-22s, or other capabilities. It is, arguably, the combination of country- and program-oriented assessments that provides the senior Air Force leadership the most comprehensive insights into the comparative contributions of all security cooperation initiatives that seek to enhance airpower and advance the security interests of the United States.

Packaging Assessments as Decision-Support Materials

Neither the Air Force nor this research team can anticipate fully all of the ways in which security cooperation assessments might be imported into Air Force management and decisionmaking.

We can, however, expect that security cooperation assessments will find a following and that various offices will make use of them.

To make security cooperation assessment information readily accessible to all interested Air Force audiences, the Air Force should consider making assessments a permanent component of SAF/IA's automated Knowledgebase system. In addition, the Air Force should publish security cooperation assessments annually (perhaps in conjunction with the Campaign Support Plan) to support management processes in the coming year. If a complete report on security cooperation assessments were available every January, such a volume could support budget and POM-related decisions, as well as force employment and force development decisions throughout the year. The composite volume should offer both country-oriented and program-oriented assessments. As discussed above, the assessments themselves need not be complicated, but would require minimal data gathering from reasonably obvious sources.

Conclusions

This chapter examined the case for the utility of enhanced security cooperation assessments: It articulated the need for a common understanding of security cooperation assessments and posited that security cooperation assessments can support DoD and USAF decisionmaking if they are available, current, accurate, and configured appropriately. The chapter offered a logic for security cooperation assessments that addressed **why** the Air Force should pursue enhanced assessments, **what** the service should assess, **how** it should go about conducting its assessments, and **who** should perform various aspects of the assessments. The chapter also applied this generic construct to the Air Force's MPEP and OET program, using data from the assessment survey described in the previous chapter.

The key to successful assessment lies, in part, with sound metrics like those suggested in this chapter. None of the individual metrics considered here should be used stochastically to drive the Air Force to specific security cooperation positions or decisions that would otherwise be considered unreasonable or ill-advised. Rather, metrics should be considered together, holistically, and over time, to reveal trends and tendencies in program and country performance. Such assessments would enable the Air Force to see clearly which countries and programs are performing positively and which ones are struggling, and to take actions as appropriate. Publishing an annual security cooperation assessments volume would support such holistic, long-term analysis and decisionmaking.

The next chapter summarizes the assessment insights and findings contained in Chapters Two, Three, and Four and offers recommendations that are intended to help the Air Force implement a comprehensive approach to assessing its efforts to build partnerships with foreign countries.

CHAPTER FIVE

Assessment Insights, Findings, and Recommendations

As articulated in the previous chapter, one major objective of this report is to strengthen the case for a comprehensive approach to security cooperation assessment by tying such an approach to the evolving Air Force discussion of security cooperation planning and execution, as well as established Air Force, joint, and interagency decisionmaking processes. To a certain extent, the rationale for this is straightforward. Neither assessments nor security cooperation are ends in and of themselves. Thus, it does not make sense to develop a free-floating assessment structure that is unconnected to security cooperation strategy and high-level departmental decisions.

That said, the need for a comprehensive security cooperation assessment approach has become more urgent in recent years. Security cooperation between U.S. defense organizations and foreign militaries has evolved into a set of events, activities, programs, and initiatives that are increasingly viewed by DoD leaders as central to shaping international relations in ways that are favorable to U.S. interests. The designation of building partnerships as an Air Force core function, in particular, suggests that it is time to end the isolation of security cooperation and integrate it more tightly with mainstream force planning and management within the Air Force, joint, and interagency realms. In part, this will require a systematic evaluation of security cooperation needs, capability packages, processes, outcomes, and cost-effectiveness: hence, the emphasis on assessment that has emerged from the suite of new security cooperation–related DoD and Air Force guidance and instructions.

As Chapter One affirms, the Air Force is experienced in the assessment realm. Many important Air Force operational and institutional activities have built-in assessment mechanisms that have proven their worth over time. These include the Red Flag Measurement and Debriefing System, Nuclear Surety Inspections, and Operational Test and Evaluation. Furthermore, the Air Force has been forging partnerships with U.S. allies and partners since before it existed as an independent service, and elements within the Air Force have become adept at performing certain kinds of security cooperation assessments.

The challenge that RAND and the Air Force are facing is how to establish an assessment approach to ensure that Air Force security cooperation programs and activities are closely aligned with operational and strategic objectives, adequately authorized and resourced, carefully sequenced and packaged, and efficiently and effectively executed. An additional hurdle to overcome is the skepticism among some in the Air Force security cooperation community regarding the need for more assessments than are currently being done, the capacity of the Air Force to perform additional assessments given existing resource constraints, and even the goal of conducting other than short-term, process-oriented assessments given the often long-term, unexpected, and indirect effects of security cooperation activities.

In 2008, RAND developed a framework for evaluating security cooperation programs and activities managed and executed by the Air Force that attempted to combine many of the theoretical, organizational, procedural, and informational strands that were necessary for focused, integrated assessment.[1] This approach included the following elements: the hierarchy of evaluation and its nested layers, the notion that assessments support decisions, a set of security cooperation stakeholder organizations, key roles in supporting assessments, and generic measures and data sources linked to the levels of the evaluation hierarchy.

RAND's 2009 study, the subject of this report, attempts to fold this approach into a larger functional and decisional context that will enable the Air Force to better meet its security cooperation assessment challenges. It does so by building on the three assumptions listed below:

1. Assessment benefits the things assessed.
2. Air Force security cooperation can benefit if there is a common understanding of security cooperation assessments.
3. Security cooperation assessments can support DoD and Air Force decisionmaking if they are available, current, accurate, and configured appropriately.

While the first assumption is affirmed in Chapter One through examples from non–security cooperation U.S. Air Force endeavors, the remainder of this chapter summarizes the insights and findings of subsequent chapters that relate to assumptions two and three and recommends ways that the Air Force might respond to these findings. It concludes by asking: How much of an effort would be required for the Air Force to implement the assessment approach recommended in this report?

Developing a Common Understanding of Security Cooperation Assessment

Comprehensive and integrated security cooperation assessments benefit the Air Force by bringing new, relevant information to bear in partnership design and prioritization and in force development and force employment, as well as by providing inputs to the PPBE process so that security cooperation can be as well managed and understood as other Air Force functions. But this can happen only if there is a common understanding among Air Force stakeholders of the rationale behind security cooperation assessment. Unfortunately, the results of our Assessment Day focus group discussions, which included Air Force security cooperation planners and executors from a variety of HAF, MAJCOM, and regional component organizations (see Chapter Two), confirmed that no consensus exists within the Air Force on four fundamental issues pertaining to security cooperation assessments: Why assess? What to assess? How to assess? And who should assess?

Consequently, we recommend that SAF/IA, as the lead within the Air Force for building partnerships with foreign countries, take the lead in developing a capstone document to guide assessment efforts across the full range of security cooperation functions, organizations, programs, and countries in which Air Force personnel operate, either as managers, implementers

---

[1]   Moroney et al., 2010a.

or observers. (See Chapter Four.) To recap, this chapter answers the four assessment questions as follows:

Why Assess?

The Air Force should conduct security cooperation assessments to improve partnership design and prioritization, force employment, force development, and overall force management decisions. Assessments of partner countries' military needs, capabilities, and objectives can all play a role in prioritizing Air Force partners and designing appropriate security cooperation programs and activities for them. In addition, assessments are necessary for making decisions on whether, how, and the extent to which the Air Force should employ personnel and units for security cooperation–related efforts based on the past performance and effectiveness of the same or similar efforts in partner countries. Assessments are also important for supplying the appropriate number and type of forces with the right kinds of skills and experience to meet the demand for security cooperation activities with foreign countries. Finally, assessments are needed to calculate the costs and benefits of security cooperation programs and engagements with different partner countries relative to other security cooperation and non–security cooperation programs and engagements.

In the case of Air Force–managed programs—such as OET and MPEP, which were the focus of our assessment survey described in Chapter Three—service officials are responsible for making decisions that relate to partnership design and prioritization, force employment, force development, and force management.

What to Assess?

The Air Force should assess security cooperation events, programs/activities, countries, and multinational initiatives. Advocates of focusing on either program or country assessments are misguided. The truth is that there cannot be a single unit of analysis for Air Force security cooperation assessments. Air Force personnel not only design and implement security cooperation programs, they also develop country plans and engage with their counterparts in particular countries. Moreover, security cooperation officials operate at different levels within the context of programs and countries. They might observe or participate in a discrete event or series of events. Alternatively, they might oversee or report on a collection of activities within a program, the program as a whole, or several programs wrapped up in a security cooperation initiative. Plus, these events, activities, programs, and initiatives can be national, multinational, regional, or even global in terms of partner involvement.

Given existing limits on resources, expertise, and authority, however, the Air Force needs to develop a step-by-step strategy for implementing a comprehensive security cooperation assessment framework. As we have argued previously, one way to do this is to focus its initial assessment efforts on the programs that they manage, such as MPEP and OET, as opposed to the programs that they execute on behalf of other government departments and DoD organizations. Thus, to use our two program examples, the Air Force should take the lead in developing methods for evaluating

- the number and type of air force–related personnel exchanges with foreign partners and the slate of countries with which the USAF leadership is engaging in operational discussions

- the satisfaction of USAF and foreign partners with the personnel exchange process and the personnel being exchanged, as well as the usefulness of operational engagement talks from a U.S. and partner perspective
- the capability of USAF air force program officials and units to service and host foreign exchange personnel and hold high-level talks with foreign military leaders
- the overall requirement for air force personnel exchanges and operational engagement talks in terms of the resources needed to match the demand.

How to Assess?

The Air Force should choose appropriate assessment types, objectives and effects, quantitative and qualitative measures, and relevant data. RAND's 2008 assessment report provides a good starting point for conducting security cooperation assessments.[2] The first step is selecting the appropriate assessment type (or level of the hierarchy of evaluation) based on the decision that the assessment is intended to inform. If the decision is related to partnership design or prioritization, then perhaps a needs assessment is in order. A force development decision might require a design/theory or process/implementation assessment. Decisions related to force employment might call for a process/implementation or an outcome/impact assessment. Cost-effectiveness assessments could be used for top-level force management decisionmaking. From there, security cooperation officials and analysts will need to apply or develop specific program- or country-oriented objectives and effects, associated quantitative or qualitative measures, and relevant sources of data.

To use our two program examples again, a needs assessment would provide information on willingness and ability of a particular country to provide and host air force exchange personnel, in the case of MPEP, and the foreign partner's desire to engage in operational discussions with the USAF and its importance to the USAF in entering into such discussion, in the case of OET. Design and theory assessments could focus on the alignment of personnel exchanges and operational engagement talks with COCOM, Air Force, and DoD strategic priorities. They would also be useful for assessing the placement of these activities within the overall package of security cooperation programs designed to enhance relationships with, build the capacity of, or enable access to particular countries. Process and implementation assessments would focus on the efficiency of MPEP and OET officials in carrying out exchanges and arranging talks, whereas outcome/impact assessments would attempt to gauge the degree to which exchanges and talks influenced the attitude or behavior of foreign air force officials in a way that contributed to service, COCOM, or DoD goals. Finally, cost-effectiveness assessments would be employed to inform security cooperation program advocates within the Air Force whether the current allocations of resources devoted to personnel exchanges and operational engagement talks were achieving the desired "bang for the buck."

Who Should Assess?

Under the joint leadership of SAF/IA and AF/A3/5, the Air Force should link department, MAJCOM/NAF, wing and below, and joint elements to appropriate roles and supporting-supported relationships. Our survey analysis of stakeholders in two very different Air Force management programs indicates that data collector and assessor roles are not always being per-

---

[2] Moroney et al., 2010a.

formed at the appropriate organizational level. Additionally, assessment information is often not being disseminated to the Air Force organization best positioned to make use of it to support security cooperation or larger departmental decisions. Thus it is imperative that SAF/IA and the Air Staff coordinate with other security cooperation stakeholders to establish general guidelines for assigning department, MAJCOM/NAF, wing and below, and joint elements, such as security assistance officers and defense attachés, to assessment roles and supporting-supported relationships within the context of different Air Force decisionmaking processes.

The results of our assessment survey suggest that Air Force–managed programs, like MPEP and OET, should have a standard set of assessment responsibilities that operate across the range of security cooperation functions, assessment focus areas, and types of assessment. In the case of MPEP (which is managed by SAF/IA), SAF/IAP, the Air Force components, and the RPMOs would coordinate efforts to collect the data necessary for all program-related assessments. The initial assessment of these data would be the responsibility of SAF/IAP in cooperation with the Air Force components. SAF/IAG would be in charge of validating and integrating these assessments with those of other Air Force security cooperation programs for higher-level decisionmaking purposes. With respect to OET, which is managed by the Air Staff, AF/A5XX and the Air Force components would take joint responsibility for data collection and initial program assessment. SAF/IA and HAF A3/5 would play the validation role, whereas SAF/IAG would undertake the integration function.

Understanding the Challenges to Effective Security Cooperation Assessment

Specific assessment benefits depend on U.S. and partner interests, security cooperation objectives, and force development requirements. For example, assessments can help OSD, the COCOMs, and the Air Force to prioritize partner countries for security cooperation funding and high-level attention based on the strategic importance of these partners, their military capacity and ability to deal with internal and external threats, their willingness to engage with the United States, and their political acceptability to the administration, Congress, and the American public. Assessments can also inform decisions regarding the establishment, continuation, expansion, or transference of programs and activities designed to build partner capacity, foster partner relationships, or guarantee U.S. access to partner countries. Additionally, assessments can provide information relevant to generating Air Force units to perform security cooperation activities, such as training additional airmen to train, advise, and assist or expanding U.S. Air Force schools to accommodate more foreign students. Yet assessments cannot provide these benefits unless they are available, current, accurate, and appropriately configured.

Although the results of our Assessment Day focus group discussions and security cooperation program survey are not definitive, they point to significant problems with the Air Force's current security cooperation assessment process. Below is a brief list of our assessment process findings. Each finding is followed by a recommended course of action for the Air Force.

Assessment Data Are Often Unavailable

SAF/IA and AF/A3/5 should take the lead in making security cooperation data collection and assessment a responsibility for Air Force personnel working in the field. They also should coordinate with joint organizations, when appropriate, to obtain information and insights needed to fully assess the performance and effectiveness of Air Force security cooperation programs.

Access to relevant data was viewed as an impediment to security cooperation assessment by most Assessment Day participants. Our survey analysis suggested why this might be such a salient issue. A significant portion of survey respondents indicated that they did report on the security cooperation activities that they observed. In particular, joint stakeholders, such as security assistance officers and defense attachés, are not being asked to share their views regarding the Air Force security cooperation events they attend.

Guidance Is Not Consistently Available or Well Understood

SAF/IA, AF/A3/5, and the MAJCOMs (especially AETC, Air Mobility Command, and the regional components) should incorporate assessment guidance in security cooperation planning documents, including specific program or country objectives, measures, and data requirements. Another finding from our survey is that program guidance, which could be used to direct assessment efforts, is either currently unavailable, as in the case of OET, or poorly understood, as in the case of MPEP. Consequently, Air Force officials attending security cooperation program events have been gathering data and evaluating events without fully understanding the objectives of the program.

Assessment Skills Could Be Improved

AETC should consider various means for improving the assessment skills of its security cooperation workforce, including a voluntary security cooperation–related schoolhouse or online assessment course. Many airmen and civilians we consulted and surveyed within the security cooperation community admitted to a lack of assessment skills, although about half of the survey respondents believed they were sufficiently skilled to perform their assessment-related duties. Of course, if the Air Force leadership does not wish to allocate resources to improve the assessment skills within the security cooperation community, it could choose to rely on the expertise of dedicated evaluation organizations—such as the Air Force Inspector General's office and HAF and MAJCOM Analyses, Assessment Studies and Lessons Learned (A9) staffs—to conduct the full range of security cooperation assessments.

Assessment Resources May Be Inadequate

SAF/IA and the Air Staff should conduct an in-depth resource analysis after determining appropriate security cooperation assessment roles and supporting-supported relationships. Assessment Day participants viewed inadequate time and resources as a major constraint to performing security cooperation assessments. However, it is not clear whether perceived resource shortfalls are derived from a realistic appraisal of actual assessment requirements or an inadequate appreciation for what is actually required to conduct assessments that support Air Force decisions.

Data Are Lacking for Effective Program Advocacy

SAF/IA and the Air Staff should ensure that data necessary for managing and advocating for security cooperation program resources are analyzed by HAF officials and used to inform PPBE recommendations and decisions. Our survey research suggests that security cooperation officials are not fully using assessments to support important managerial and budgetary decisions. With respect to the two programs we examined, relatively few departmental stakeholders claimed they managed or advocated for program resources, apparently relying on MAJCOM/NAF and joint stakeholders to perform these functions. Presumably, this situation would change if HAF policymakers began to demand that security cooperation program recommendations be supported by reliable and complete information on cost, performance, effectiveness, and unfulfilled requirements.

Officials Are Unable to Compare Programs

SAF/IA, AF/A3/5, and AETC should work together to educate airmen and civilians on the broader universe of Air Force security cooperation by creating a handbook of Air Force programs and activities that are designed to build partnerships with foreign countries. Alternatively, the Air Force should incorporate this information into the Air Force Campaign Support Plan as an appendix. Survey respondents demonstrated little insight at any organizational level as to how their programs compared with other similar security cooperation programs, making it difficult to provide decisionmakers with cost-effectiveness information.

Implementing a New Approach to Security Cooperation Assessment

Establishing a comprehensive and integrated approach to security cooperation assessment should not be an extraordinary burden for the Air Force. Much of the information necessary for security cooperation assessment is available, or could be available, if the Air Force ordered its collection and managed its dissemination, analysis, and integration into Air Force and DoD decisionmaking processes. The authorities seem to exist, and additional Air Force instructions could fill in certain gaps. The Air Force has forged cooperative relationships across disparate communities in the past when prudence dictated; there is no reason to believe similar relationships could not be developed in support of security cooperation assessment.

However, there are certain hurdles impeding the establishment of a new approach to security cooperation assessment. These include the doctrinal and organizational (i.e., why, what, how, who) issues summarized in the first section of this chapter, as well as the procedural, educational, and resource issues discussed in the second section. The hurdles also include an analytical issue and a policy issue.

Those responsible for security cooperation in the Air Force need to more fully understand the dimensions of the assessment capacity problem and communicate capacity requirements to senior Air Force and DoD leaders. Toward this end, we recommend that AF/A3/5 and SAF/IA continue to survey the full range of participants in Air Force security cooperation programs regarding their current assessment-related roles, responsibilities, resources, data sources, guidance, and skills. In addition, Air Force security cooperation officials need to decide on a strategy for implementing a new assessment approach. We suggest the following four-part strategy:

1. Achieve a consensus among security cooperation stakeholders on the elements of the overall assessment approach, taking into account the best practices employed by other services and defense agencies; other government agencies, such as the State Department and the U.S. Agency for International Development; and major allies, such as the United Kingdom and France.
2. Enshrine this vision in a stand-alone instruction or as an annex to the Air Force Campaign Support Plan or the Air Force Global Partnership Strategy.

3. Gradually implement the new assessment approach, focusing first on Air Force–managed programs and activities that are well established, clearly defined, and adequately resourced.
4. Actively collaborate with other DoD and Department of State stakeholders regarding assessment policy affecting security cooperation programs and activities in which the Air Force participates but does not manage.

The question is no longer whether the Air Force should assess its efforts to build partnerships. Rather, it is how to do so in an integrated, focused way that takes into account the evolving nature of security cooperation policy as well as Air Force constraints with respect to assessment capacity and authority. Ongoing discussions related to the development of the Campaign Support Plan and the Building Partnerships Core Function Master Plan provide near-term opportunities for achieving and executing a consensus within the Air Force regarding security cooperation assessment.

APPENDIX A
Assessment Survey Approach and Results

This appendix begins by describing the approach used in developing our assessment survey of Air Force security cooperation program stakeholders. Next, it provides a listing of the questions used to elicit responses regarding assessment roles and data types. To help the reader better understand the analytical steps used by RAND, the questions are grouped according to their relation to assessment roles and data types, as opposed to the order in which they were presented to the survey respondents. (Appendix B provides the survey in its entirety.) Finally, the appendix provides a series of tables that show the results of the survey using the analytical construct.

Survey Development

In developing the survey instrument, the RAND study team made several assumptions regarding the types of data that would be useful in conducting assessments. First, as described in Chapter Three, the study team postulated five broad types of data, including demand, throughput, resources, cost, and objectives. The first two of these are related, as are the second two types of data. Demand data, such as requests to participate, are closely linked to throughput data, which refers to data regarding actual participation. Resource data, such as how much funding or manpower might be available to implement the program, are closely linked to cost data, which account for actual expenditures during program implementation. By objectives, we mean data regarding indications of how well the program might be achieving its overall outcome objectives as well as how well it might be achieving its activity output objectives. Specific examples of these data were drawn from AFIs and similar guidance documents, as depicted in Table 3.1, and repeated here.

Second, certain types of data might support specific types of assessments. The data that would be relevant to a cost-effectiveness assessment, for example, might be different from the data necessary for assessing a program's design and theory. Table A.1 shows the five types of assessments and the types of data used to support them.

These relations were a key element in designing the survey, as many of the survey questions attempt to determine whether a respondent has access to specific types of data. Understanding these relationships allowed the study team to determine which respondents' organizations could potentially collect data and how that data could be used to support specific types of assessments. Accordingly, the survey also included questions that attempted to identify respondents' organizations that might be in a position to use the data to actually conduct an assessment.

57

Table A.1
Data Supports Security Cooperation Assessments

Type of Assessment	Type of Data				
	Demand	Throughput	Resources	Cost	Objectives
Need for program	X			X	X
Design and theory	X		X		X
Process/implementation	X		X		X
Outcome/impact	X				X
Cost-effectiveness		X	X	X	X

NOTE: The types of assessment are those associated with RAND's hierarchy of evaluation. The types of data are those deemed necessary to support specific types of assessment.

The survey respondents did not see the questions presented this way; rather, they were organized and presented in an order that followed the current task construct described in Chapter Four. The study team believed the survey would be more easily understood and would, in turn, contain more accurate responses if we asked questions about the respondents' current duties as opposed to placing them in a position of having to answer hypothetical questions about how they might conduct assessments. To do this, we developed four roles that a respondent might have in support of a program: (1) Implementing a Program, (2) Designing a Program, (3) Making Program Recommendations, and (4) Making Program Decisions. While making the survey questions understandable and relevant for the respondent, the questions were also designed to allow the study to infer potential assessment roles and to understand the types of data the respondents' organizations used or had access to.

The following section lists each of these questions, grouped by the type of data they support, as shown in Table A.1. To help the reader understand the actual presentation order of the questions as they appeared to the respondents, the question number as it appeared in the survey is included.

Survey Questions

Demand
Potential Data Collector

11. Do you collect data regarding requests to participate in the ?

    ○   No
    ○   Don't know/Not applicable
    ○   Yes

12. If you answered "Yes" to the previous question, please describe the types of data you collect and the sources of the data:

    ○   Country List
    ○   Country Nomination

- ○ Guest Lists
- ○ International Visit Request
- ○ Invitation Travel Orders
- ○ Letter of Request
- ○ My own observations/trip report
- ○ Nomination Package
- ○ Program Request
- ○ Project Agreement/Arrangement
- ○ Project Nomination Form
- ○ Project Proposal
- ○ Proposal for PME Exchange
- ○ Request for Air Force Personnel Attendance
- ○ Request or Use of Air Force Aircraft
- ○ Summary of Statement of Intent
- ○ Training Quotas
- ○ Travel Orders
- ○ Visit Request
- ○ Other, please specify

17. Do you collect data regarding capabilities assessments for the ?

- ○ No
- ○ Don't know/Not applicable
- ○ Yes

18. If you answered "Yes" to the previous question, please describe the types of data you collect and the sources of the data:

- ○ Exchange Agreement
- ○ Interim Tour Report
- ○ International Agreement
- ○ Memorandum of Agreement
- ○ Memorandum of Understanding
- ○ My own observations/trip report
- ○ Participant Entry or Exit Testing
- ○ Quid-Pro-Quo Analysis
- ○ Summary of Statement of Intent
- ○ Other, please specify

Potential Assessor

29. Do you design, or contribute to the design of, specific events or activities for the ?

- ○ No
- ○ Don't know/Not applicable
- ○ Yes

30. If you answered "Yes" to the previous question, please list the events or activities that you design, or contribute to the design of:

- O   Agenda
- O   Background Material
- O   Briefings
- O   Budget Allocation Memo
- O   Budget Projection
- O   Budget Request
- O   Country List
- O   Country Nomination
- O   Exchange Agreement
- O   Guest Lists
- O   Master Data Agreement
- O   Master Information Agreement
- O   Position Description and Requisition Report
- O   Program Request
- O   Project Proposal
- O   Proposal for PME Exchange
- O   Request for Disclosure Authorization
- O   Request for Fund Cite
- O   Request for Use of Air Force Aircraft
- O   Request for Air Force Personnel Attendance
- O   Security Plan
- O   Summary Statement of Intent
- O   Test and Evaluation Plan
- O   Visit Request
- O   Other, please specify

39. Do you make recommendations regarding the overall need for the ?

- O   No
- O   Don't know/Not applicable
- O   Yes, please list the stakeholder(s) (i.e., office or organization) that receive(s) your recommendations:

40. Do you make recommendations regarding the need to increase or reduce participation in the ?

- O   No
- O   Don't know/Not applicable
- O   Yes, please list the stakeholder(s) (i.e., office or organization) that receive(s) your recommendations:

41. Do you make recommendations regarding which countries participate in the ?

- ○ No
- ○ Don't know/Not applicable
- ○ Yes, please list the stakeholder(s) (i.e., office or organization) that receive(s) your recommendations:

Throughput
Potential Data Collector

13. Do you collect data regarding attendees, participants, or numbers of graduates for the ?

- ○ No
- ○ Don't know/Not applicable
- ○ Yes

14. If you answered "Yes" to the previous question, please describe the types of data you collect and the sources of the data:

- ○ Country List
- ○ Country Nomination
- ○ Guest Lists
- ○ International Visit Request
- ○ Invitational Travel Orders
- ○ Letter of Acceptance
- ○ Letter of Request
- ○ My own observations/trip report
- ○ Nomination Package
- ○ Proposal for PME Exchange
- ○ Request for Air Force Personnel Attendance
- ○ Training Quotas
- ○ Travel Orders
- ○ Visit Request
- ○ Other, please specify

Potential Assessor

26. Do you prepare reports that document specific activities or events (i.e., trip reports, after-action reports, surveys, etc.) for the ?

- ○ No
- ○ Don't know/Not applicable
- ○ Yes

27. If you answered "Yes" to the previous question, please identify these documents:

- ○ After-Action Report

○   Annual Report
○   End of Tour Report
○   Interim Tour Report
○   My own observations/trip report
○   Progress Report
○   Project Final Report
○   Quarterly Obligation Report
○   Test and Disposition Report
○   Training Report
○   Other, please specify

Resources
Potential Data Collector

7. Do you manage resources that are used in the implementation of the ?

○   No
○   Don't know/Not applicable
○   Yes

8. If you answered "Yes" to the previous question, please list the resources you manage:

○   Facilities
○   Funds
○   Infrastructure
○   People
○   Other, please specify

Potential Assessor

29. Do you design, or contribute to the design of, specific events or activities for the ?

○   No
○   Don't know/Not applicable
○   Yes

30. If you answered "Yes" to the previous question, please list the events or activities that you design, or contribute to the design of:

○   Agenda
○   Background Material
○   Briefings
○   Budget Allocation Memo
○   Budget Projection
○   Budget Request
○   Country List
○   Country Nomination

○ Exchange Agreement
○ Guest Lists
○ Master Data Agreement
○ Master Information Agreement
○ Position Description and Requisition Report
○ Program Request
○ Project Proposal
○ Proposal for PME Exchange
○ Request for Disclosure Authorization
○ Request for Fund Cite
○ Request for Use of Air Force Aircraft
○ Request for Air Force Personnel Attendance
○ Security Plan
○ Summary Statement of Intent
○ Test and Evaluation Plan
○ Visit Request
○ Other, please specify

44. Do you advocate for funds used to implement the ?

○ No
○ Don't know/Not applicable
○ Yes, please list the process(es) you use (i.e., PPBE, requests for O&M, ORF, or other existing funding sources, etc.):

Cost
Potential Data Collector

21. Do you collect data regarding the funds expended for the ?

○ No
○ Don't know/Not applicable
○ Yes

22. If you answered "Yes" to the previous question, please describe the types of data you collect and the sources of the data:

○ Budget Allocation Memo
○ Budget Projection
○ Budget Request
○ Loan Agreement
○ My own observations/trip report
○ Periodic Financial Report
○ Quarterly Obligation Report
○ Request for Fund Cite
○ Travel Vouchers
○ Other, please specify

26. Do you prepare reports that document specific activities or events (i.e., trip reports, after-action reports, surveys, etc.) for the ?

   ○ No
   ○ Don't know/Not applicable
   ○ Yes

27. If you answered "Yes" to the previous question, please identify these documents:

   ○ After-Action Report
   ○ Annual Report
   ○ End of Tour Report
   ○ Interim Tour Report
   ○ My own observations/trip report
   ○ Progress Report
   ○ Project Final Report
   ○ Quarterly Obligation Report
   ○ Test and Disposition Report
   ○ Training Report
   ○ Other, please specify

36. Do you collect data regarding cost of the overall program or the cost of a unit of output (i.e., one graduate, one event, etc.)?

   ○ No
   ○ Don't know/Not applicable
   ○ Yes

37. If you answered "Yes" to the previous question, please identify this data:

   ○ Budget Allocation Memo
   ○ Budget Projection
   ○ Budget Request
   ○ Loan Agreement
   ○ Periodic Financial Report
   ○ Quarterly Obligation Report
   ○ Request for Fund Cite
   ○ Travel Orders
   ○ Travel Vouchers
   ○ Other, please specify

Potential Assessor

48. Do you have access to information regarding other USAF programs' security cooperation programs, such as their objectives, cost, and benefits?

       ○   No
       ○   Don't know/Not applicable
       ○   Yes, what are the sources for this information?

49. Do you have access to information regarding other USAF programs (not security cooperation) and the priority attached to each?

       ○   No
       ○   Don't know/Not applicable
       ○   Yes, what are your sources for this information?

50. Do you have access to information regarding other USAF programs (not security cooperation) and their objectives, cost, and benefits?

       ○   No
       ○   Don't know/Not applicable
       ○   Yes, what are your sources for this information?

Objectives
Potential Data Collector

9. Do you directly observe or participate in any of the specific events or activities related to the ?

       ○   No
       ○   Don't know/Not applicable
       ○   Yes

10. If you answered "Yes" to the previous question, please list the events or activities in which you observe or participate:

       ○   Command Post Exercise
       ○   Competition
       ○   Computer-Assisted Simulation or Gaming Activity
       ○   Conference or Roundtable Discussion
       ○   Education
       ○   Field Exercise
       ○   Personnel Exchange
       ○   Table-Top Exercise
       ○   Test of Field Experiment
       ○   Training
       ○   Other, please specify

15. Do you collect data regarding participant views or observations, such as exit surveys for the ?

- ○ No
- ○ Don't know/Not applicable
- ○ Yes

16. If you answered "Yes" to the previous question, please describe the types of data you collect and the sources of the data:

- ○ After-Action Report
- ○ Annual Report
- ○ End of Tour Report
- ○ Meeting Minutes / Summary
- ○ My own observations/trip report
- ○ Progress Report
- ○ Project Final Report
- ○ Project Quarterly Report
- ○ Test and Disposition Report
- ○ Training Report
- ○ Other, please specify

19. Do you collect data regarding the effect of the ? on relevant partner capabilities?

- ○ No
- ○ Don't know/Not applicable
- ○ Yes

20. If you answered "Yes" to the previous question, please describe the types of data you collect and the sources of the data:

- ○ After-Action Report
- ○ Alumni Whereabouts
- ○ Annual Report
- ○ Certification to Congress
- ○ End of Tour Report
- ○ Interim Tour Report
- ○ Meeting Minutes/Summary
- ○ My own observations/trip report
- ○ Progress Report
- ○ Project Quarterly Report
- ○ Quid-Pro-Quo Analysis
- ○ Test and Disposition Report
- ○ Training Report
- ○ Other, please specify

23. Do you collect data regarding compliance with legal or regulatory requirements related to the ?

- ○ No
- ○ Don't know/Not applicable
- ○ Yes

24. If you answered "Yes" to the previous question, please describe the types of data you collect and the sources of the data:

- ○ Data Exchange Annex
- ○ Delegation of Disclosure Authority Letter
- ○ Extended Visit Authorization
- ○ Information Exchange Agreement
- ○ My own observations/trip report
- ○ Request for Disclosure Authorization
- ○ Security Plan
- ○ Other, please specify

Potential Assessor

26. Do you prepare reports that document specific activities or events (i.e., trip reports, after-action reports, surveys, etc.) for the ?

- ○ No
- ○ Don't know/Not applicable
- ○ Yes

27. If you answered "Yes" to the previous question, please identify these documents:

- ○ After-Action Report
- ○ Annual Report
- ○ End of Tour Report
- ○ Interim Tour Report
- ○ My own observations/trip report
- ○ Progress Report
- ○ Project Final Report
- ○ Quarterly Obligation Report
- ○ Test and Disposition Report
- ○ Training Report
- ○ Other, please specify

42. Do you gather data for the ? that reflects how well a specific event or activity met its objectives?

- ○ No
- ○ Don't know/Not applicable
- ○ Yes

43. If you answered "Yes" to the previous question, please describe the type of data:

        O   After-Action Report
        O   Annual Report
        O   Certification to Congress
        O   End of Tour Report
        O   Interim Tour Report
        O   Meeting Minutes/Summary
        O   My own observations/trip report
        O   Progress Report
        O   Project Final Report
        O   Project Quarterly Report
        O   Test and Disposition Report
        O   Training Report
        O   Quarterly Obligation Report
        O   Other, please specify

46. Do you set the objectives for the overall program?

        O   No
        O   Don't know/Not applicable
        O   Yes, please indicate where the objectives are documented

47. Do you set the objectives for specific events or activities within the ?

        O   No
        O   Don't know/Not applicable
        O   Yes, please indicate where the objectives are documented

Survey Results

This section provides a full set of tables showing the results of responses for both OET and MPEP. These tables are explained in detail in Chapter Three; Tables 3.3 and 3.4 are drawn from this set of tables. Tables A.2 and A.3 summarize the results of analyzing the groups of questions relating to potential data collector and assessor roles for the various types of assessments. On the left side of each table are the types of data and the associated questions, and on the right side are the results of the questions related to data collecting and assessing. The percentages in each of the cells represent the proportion of respondents indicating they collect the subject type of information, or perform a function that suggests they could perform an assessment role. Table A.2 summarizes the results of the analysis for OET. Three percentages are displayed in each cell. These percentages correspond, respectively, to department-level respondents, MAJCOM and NAF respondents, and joint respondents. Table A.3 shows the results for MPEP; it includes a fourth level of respondent: wing and below.

Table A.2
Operator Engagement Talks Assessment Roles Results, by Role and Level

Data Type	Associated Questions	Data Collector			Assessor		
		Dept.	MAJCOM or NAF	Joint	Dept.	MAJCOM or NAF	Joint
Operator Engagement Talks—Cost-Effectiveness							
Throughput	13. Do you collect data regarding attendees, participants, or numbers of graduates for the ?	43	25	75			
	26. Do you prepare reports that document specific activities or events (i.e., trip reports, after-action reports, surveys, etc.) for the ?				57	50	75
Resources	7. Do you manage resources that are used in the implementation of the ?	43	50	25			
	29. Do you design, or contribute to the design of, specific events or activities for the ?				43	75	75
	44. Do you advocate for funds used to implement the ?				0	0	25
Cost	21. Do you collect data regarding the funds expended for the ?	14	50	0			
	26. Do you prepare reports that document specific activities or events (i.e., trip reports, after-action reports, surveys, etc.) for the ?				57	50	75
	36. Do you collect data regarding cost of the overall program or the cost of a unit of output (i.e., one graduate, one event, etc.)?	0	0	0			
	48. Do you have access to information regarding other USAF programs' security cooperation programs, such as their objectives, cost, and benefits?				14	25	0
	49. Do you have access to information regarding other USAF programs (not security cooperation) and the priority attached to each?				0	50	25
	50. Do you have access to information regarding other USAF programs (not security cooperation) and their objectives, cost, and benefits?				0	50	0

Table A.2—Continued

Data Type	Associated Questions	Data Collector			Assessor		
		Dept.	MAJCOM or NAF	Joint	Dept.	MAJCOM or NAF	Joint
Objectives	9. Do you directly observe or participate in any of the specific events or activities related to the ?	72	75	75			
	15. Do you collect data regarding participant views or observations, such as exit surveys for the ?	43	75	75			
	19. Do you collect data regarding the effect of the ? on relevant partner capabilities?	14	25	50			
	26. Do you prepare reports that document specific activities or events (i.e., trip reports, after-action reports, surveys, etc.) for the ?				57	50	75
	42. Do you gather data for the ? that reflects how well a specific event or activity met its objectives?				29	50	50
	46. Do you set the objectives for the overall program?				0	0	25
	47. Do you set the objectives for specific events or activities within the ?				14	0	0
Operator Engagement Talks—Outcome/Impact							
Demand	11. Do you collect data regarding requests to participate in the ?	43	50	50			
	17. Do you collect data regarding capabilities assessments for the ?	29	50	50			
	29. Do you design, or contribute to the design of, specific events or activities for the ?				43	75	75
Objectives	9. Do you directly observe or participate in any of the specific events or activities related to the ?	72	75	75			
	15. Do you collect data regarding participant views or observations, such as exit surveys for the ?	43	75	75			
	19. Do you collect data regarding the effect of the ? on relevant partner capabilities?	14	25	50			
	26. Do you prepare reports that document specific activities or events (i.e., trip reports, after-action reports, surveys, etc.) for the ?				57	50	75
	42. Do you gather data for the ? that reflects how well a specific event or activity met its objectives?				29	50	50
	46. Do you set the objectives for the overall program?				0	0	25
	47. Do you set the objectives for specific events or activities within the ?				14	0	0

Table A.2—Continued

Data Type	Associated Questions	Data Collector			Assessor		
		Dept.	MAJCOM or NAF	Joint	Dept.	MAJCOM or NAF	Joint
Operator Engagement Talks—Process/Implementation							
Demand	11. Do you collect data regarding requests to participate in the ?	43	50	50			
	17. Do you collect data regarding capabilities assessments for the ?	29	50	50			
	29. Do you design, or contribute to the design of, specific events or activities for the ?				43	75	7
Resources	7. Do you manage resources that are used in the implementation of the ?	43	50	25			
	29. Do you design, or contribute to the design of, specific events or activities for the ?				43	75	75
	44. Do you advocate for funds used to implement the ?				0	0	25
Objectives	9. Do you directly observe or participate in any of the specific events or activities related to the ?	72	75	75			
	15. Do you collect data regarding participant views or observations, such as exit surveys for the ?	43	75	75			
	19. Do you collect data regarding the effect of the ? on relevant partner capabilities?	14	25	50			
	23. Do you collect data regarding compliance with legal or regulatory requirements related to the ?	29	25	25			
	26. Do you prepare reports that document specific activities or events (i.e., trip reports, after-action reports, surveys, etc.) for the ?				57	50	75
	42. Do you gather data for the ? that reflects how well a specific event or activity met its objectives?				29	50	50
	46. Do you set the objectives for the overall program?				0	0	25
	47. Do you set the objectives for specific events or activities within the ?				14	0	0

Table A.2—Continued

Data Type	Associated Questions	Data Collector			Assessor		
		Dept.	MAJCOM or NAF	Joint	Dept.	MAJCOM or NAF	Joint
Operator Engagement Talks—Design and Theory							
Demand	11. Do you collect data regarding requests to participate in the ?	43	50	50			
	17. Do you collect data regarding capabilities assessments for the ?	29	50	50			
	29. Do you design, or contribute to the design of, specific events or activities for the ?				43	75	75
	40. Do you make recommendations regarding the need to increase or reduce participation in the ?				29	50	25
	41. Do you make recommendations regarding which countries participate in the ?				0	25	25
Resources	7. Do you manage resources that are used in the implementation of the ?	43	50	25			
	29. Do you design, or contribute to the design of, specific events or activities for the ?				43	75	75
	44. Do you advocate for funds used to implement the ?				0	0	25
Objectives	9. Do you directly observe or participate in any of the specific events or activities related to the ?	72	75	75			
	15. Do you collect data regarding participant views or observations, such as exit surveys for the ?	43	75	75			
	19. Do you collect data regarding the effect of the ? on relevant partner capabilities?	14	25	50			
	23. Do you collect data regarding compliance with legal or regulatory requirements related to the ?	29	25	25			
	26. Do you prepare reports that document specific activities or events (i.e., trip reports, after-action reports, surveys, etc.) for the ?				57	50	75
	42. Do you gather data for the ? that reflects how well a specific event or activity met its objectives?				29	50	50
	46. Do you set the objectives for the overall program?				0	0	25
	47. Do you set the objectives for specific events or activities within the ?				14	0	0

Table A.2—Continued

Data Type	Associated Questions	Data Collector Dept.	Data Collector MAJCOM or NAF	Data Collector Joint	Assessor Dept.	Assessor MAJCOM or NAF	Assessor Joint
Operator Engagement Talks—Need for Program							
Demand	11. Do you collect data regarding requests to participate in the ?	43	50	50			
	17. Do you collect data regarding capabilities assessments for the ?	29	50	50			
	29. Do you design, or contribute to the design of, specific events or activities for the ?				43	75	75
	39. Do you make recommendations regarding the overall need for the ?				0	50	75
	45. Do you determine the overall need for the ?				0	0	25
Cost	21. Do you collect data regarding the funds expended for the ?	14	50	0			
	26. Do you prepare reports that document specific activities or events (i.e., trip reports, after-action reports, surveys, etc.) for the ?				57	50	75
	36. Do you collect data regarding cost of the overall program or the cost of a unit of output (i.e., one graduate, one event, etc.)?	0	0	0			
	48. Do you have access to information regarding other USAF programs' security cooperation programs, such as their objectives, cost, and benefits?	14	25		14	25	0
	49. Do you have access to information regarding other USAF programs (not security cooperation) and the priority attached to each?				0	50	25
	50. Do you have access to information regarding other USAF programs (not security cooperation) and their objectives, cost, and benefits?				0	50	0
Objectives	9. Do you directly observe or participate in any of the specific events or activities related to the ?	72	75	75			
	15. Do you collect data regarding participant views or observations, such as exit surveys for the ?	43	75	75			
	19. Do you collect data regarding the effect of the ? on relevant partner capabilities?	14	25	50			
	23. Do you collect data regarding compliance with legal or regulatory requirements related to the ?	29	25	25			
	42. Do you gather data for the ? that reflects how well a specific event or activity met its objectives?				29	50	50
	46. Do you set the objectives for the overall program?				0	0	25
	47. Do you set the objectives for specific events or activities within the ?				14	0	0

Table A.3
Military Personnel Exchange Program Assessment Roles Results, by Role and Level

Data Type	Associated Questions	Data Collector				Assessor			
		Dept.	MAJCOM or NAF	Joint	Wing and Below	Dept.	MAJCOM or NAF	Joint	Wing and Below
Military Personnel Exchange Program—Cost-Effectiveness									
Throughput	13. Do you collect data regarding attendees, participants, or numbers of graduates for the ?	100	50	16	43				
	26. Do you prepare reports that document specific activities or events (i.e., trip reports, after-action reports, surveys, etc.) for the ?					100	42	74	14
Resources	7. Do you manage resources that are used in the implementation of the ?	100	75	20	57				
	29. Do you design, or contribute to the design of, specific events or activities for the ?					100	25	30	43
	44. Do you advocate for funds used to implement the ?					0	17	6	14
Cost	21. Do you collect data regarding the funds expended for the ?	100	42	24	29				
	26. Do you prepare reports that document specific activities or events (i.e., trip reports, after-action reports, surveys, etc.) for the ?					100	42	74	14
	36. Do you collect data regarding cost of the overall program or the cost of a unit of output (i.e., one graduate, one event, etc.)?	100	8	10	43				
	48. Do you have access to information regarding other USAF programs' security cooperation programs, such as their objectives, cost, and benefits?					0	17	2	43
	49. Do you have access to information regarding other USAF programs (not security cooperation) and the priority attached to each?					0	0	0	14
	50. Do you have access to information regarding other USAF programs (not security cooperation) and their objectives, cost, and benefits?					0	0	2	0

Table A.3—Continued

Data Type	Associated Questions	Data Collector				Assessor			
		Dept.	MAJCOM or NAF	Joint	Wing and Below	Dept.	MAJCOM or NAF	Joint	Wing and Below
Objectives	9. Do you directly observe or participate in any of the specific events or activities related to the ?	100	83	88	72				
	15. Do you collect data regarding participant views or observations, such as exit surveys for the ?	100	67	34	29				
	19. Do you collect data regarding the effect of the ? on relevant partner capabilities?	100	17	40	29				
	26. Do you prepare reports that document specific activities or events (i.e., trip reports, after-action reports, surveys, etc.) for the ?					100	42	74	14
	42. Do you gather data for the ? that reflects how well a specific event or activity met its objectives?					100	8	30	14
	46. Do you set the objectives for the overall program?					0	0	0	14
	47. Do you set the objectives for specific events or activities within the ?					0	8	8	0
Military Personnel Exchange Program—Outcome/Impact									
Demand	11. Do you collect data regarding requests to participate in the ?	100	67	32	86				
	17. Do you collect data regarding capabilities assessments for the ?	100	25	30	43				
	29. Do you design, or contribute to the design of, specific events or activities for the ?					100	25	30	43

Table A.3—Continued

Data Type	Associated Questions	Data Collector				Assessor			
		Dept.	MAJCOM or NAF	Joint	Wing and Below	Dept.	MAJCOM or NAF	Joint	Wing and Below
Objectives	9. Do you directly observe or participate in any of the specific events or activities related to the ?	100	83	88	72				
	15. Do you collect data regarding participant views or observations, such as exit surveys for the ?	100	67	34	29				
	19. Do you collect data regarding the effect of the ? on relevant partner capabilities?	100	17	40	29				
	26. Do you prepare reports that document specific activities or events (i.e., trip reports, after-action reports, surveys, etc.) for the ?					100	42	74	14
	42. Do you gather data for the ? that reflects how well a specific event or activity met its objectives?					100	8	30	14
	46. Do you set the objectives for the overall program?					0	0	0	14
	47. Do you set the objectives for specific events or activities within the ?					0	8	8	0
Military Personnel Exchange Program—Process/Implementation									
Demand	11. Do you collect data regarding requests to participate in the ?	100	67	32	86				
	17. Do you collect data regarding capabilities assessments for the ?	100	25	30	43				
	29. Do you design, or contribute to the design of, specific events or activities for the ?					100	25	30	43
Resources	7. Do you manage resources that are used in the implementation of the ?	100	75	20	57				
	29. Do you design, or contribute to the design of, specific events or activities for the ?					100	25	30	43
	44. Do you advocate for funds used to implement the ?					0	17	6	14

Table A.3—Continued

Data Type	Associated Questions	Data Collector				Assessor			
		Dept.	MAJCOM or NAF	Joint	Wing and Below	Dept.	MAJCOM or NAF	Joint	Wing and Below
Objectives	9. Do you directly observe or participate in any of the specific events or activities related to the ?	100	83	88	72				
	15. Do you collect data regarding participant views or observations, such as exit surveys for the ?	100	67	34	29				
	19. Do you collect data regarding the effect of the ? on relevant partner capabilities?	100	17	40	29				
	23. Do you collect data regarding compliance with legal or regulatory requirements related to the ?	100	50	14	29				
	26. Do you prepare reports that document specific activities or events (i.e., trip reports, after-action reports, surveys, etc.) for the ?					100	42	74	14
	42. Do you gather data for the ? that reflects how well a specific event or activity met its objectives?					100	8	30	14
	46. Do you set the objectives for the overall program?					0	0	0	14
	47. Do you set the objectives for specific events or activities within the ?					0	8	8	0
Military Personnel Exchange Program—Design and Theory									
Demand	11. Do you collect data regarding requests to participate in the ?	100	67	32	86				
	17. Do you collect data regarding capabilities assessments for the ?	100	25	30	43				
	29. Do you design, or contribute to the design of, specific events or activities for the ?					100	25	30	43
	40. Do you make recommendations regarding the need to increase or reduce participation in the ?					0	42	28	23
	41. Do you make recommendations regarding which countries participate in the ?					0	33	6	29

Table A.3—Continued

Data Type	Associated Questions	Data Collector				Assessor			
		Dept.	MAJCOM or NAF	Joint	Wing and Below	Dept.	MAJCOM or NAF	Joint	Wing and Below
Resources	7. Do you manage resources that are used in the implementation of the ?	100	75	20	57				
	29. Do you design, or contribute to the design of, specific events or activities for the ?					100	25	30	43
	44. Do you advocate for funds used to implement the ?					0	17	6	14
Objectives	9. Do you directly observe or participate in any of the specific events or activities related to the ?	100	83	88	72				
	15. Do you collect data regarding participant views or observations, such as exit surveys for the ?	100	67	34	29				
	19. Do you collect data regarding the effect of the ? on relevant partner capabilities?	100	17	40	29				
	23. Do you collect data regarding compliance with legal or regulatory requirements related to the ?	100	50	14	29				
	26. Do you prepare reports that document specific activities or events (i.e., trip reports, after-action reports, surveys, etc.) for the ?					100	42	74	14
	42. Do you gather data for the ? that reflects how well a specific event or activity met its objectives?					100	8	30	14
	46. Do you set the objectives for the overall program?					0	0	0	14
	47. Do you set the objectives for specific events or activities within the ?					0	8	8	0
Military Personnel Exchange Program—Need for Program									
Demand	11. Do you collect data regarding requests to participate in the ?	100	67	32	86				
	17. Do you collect data regarding capabilities assessments for the ?	100	25	30	43				
	29. Do you design, or contribute to the design of, specific events or activities for the ?					100	25	30	43
	39. Do you make recommendations regarding the overall need for the ?					0	50	36	43
	45. Do you determine the overall need for the ?					0	0	2	14

Table A.3—Continued

Data Type	Associated Questions	Data Collector				Assessor			
		Dept.	MAJCOM or NAF	Joint	Wing and Below	Dept.	MAJCOM or NAF	Joint	Wing and Below
Cost	21. Do you collect data regarding the funds expended for the ?	100	42	24	29				14
	26. Do you prepare reports that document specific activities or events (i.e., trip reports, after-action reports, surveys, etc.) for the ?					100	42	74	14
	36. Do you collect data regarding cost of the overall program or the cost of a unit of output (i.e., one graduate, one event, etc.)?	100	8	10	43				
	48. Do you have access to information regarding other USAF programs' security cooperation programs, such as their objectives, cost, and benefits?					0	17	2	43
	49. Do you have access to information regarding other USAF programs (not security cooperation) and the priority attached to each?					0	0	0	14
	50. Do you have access to information regarding other USAF programs (not security cooperation) and their objectives, cost, and benefits?					0	0	2	0
Objectives	9. Do you directly observe or participate in any of the specific events or activities related to the ?	100	83	88	72				
	15. Do you collect data regarding participant views or observations, such as exit surveys for the ?	100	67	34	29				
	19. Do you collect data regarding the effect of the ? on relevant partner capabilities?	100	17	40	29				
	23. Do you collect data regarding compliance with legal or regulatory requirements related to the ?	100	50	14	29				
	42. Do you gather data for the ? that reflects how well a specific event or activity met its objectives?					100	8	30	14
	46. Do you set the objectives for the overall program?					0	0	0	14
	47. Do you set the objectives for specific events or activities within the ?					0	8	8	0

Data Collection Results for OET and MPEP

Table A.4 presents the results of our survey of MPEP and OET stakeholder officials by the number of responses associated each organizational level or type: i.e., department, MAJCOM or NAF, wing and below, and joint.

Table A.4
Number of Survey Responses, by Program and Organizational Level/Type

Questions/Specific Types of Data	Military Personnel Exchange Program				Operator Engagement Talks		
	Dept.	MAJCOM or NAF	Joint	Wing and Below	Dept.	MAJCOM or NAF	Joint
Question 7: Do you manage resources that are used in the implementation of the ?							
Question 8: If you answered "Yes" to the previous question, please list the resources you manage:							
Facilities			2	5		2	
Funds	1	3	3	8		3	
Infrastructure			2	6		2	
People	1	8	3	9	1	3	1
Question 9: Do you directly observe or participate in any of the specific events or activities related to the ?							
Question 10: If you answered "Yes" to the previous question, please list the events or activities:							
Command Post Exercise			1	4			1
Competition		1		6			
Computer-Assisted Simulation or Gaming Activity		1	1	6			
Conference or Roundtable Discussion	1	5	1	15	3	4	3
Education	1	4	1	16	1		1
Field Exercise		1	1	3		1	1
Personnel Exchange	1	8	5	39		1	
Table-Top Exercise			1	3		1	
Test of Field Experiment			1	4			
Training	1	3	2	20	1	1	1
Question 11: Do you collect data regarding requests to participate in the ?							
Question 12: If you answered "Yes" to the previous question, please describe the types of data you collect and the sources of the data:							
Country List	1	2	2	2		1	
Country Nomination	1	5	4	2		1	
Guest Lists		1	2		1	1	1
International Visit Request		6	2	3	1	2	2
Invitation Travel Orders		4	3	2			
Letter of Request		4	3	3		1	1
My own observations/trip report	1	4	2	12	2	2	2
Nomination Package	1	6	3	6			
Program Request		3	2	1		2	1
Project Agreement/Arrangement	1	1	2	2	1	2	2
Project Nomination Form		1	2				
Project Proposal		1	2	1		1	1
Proposal for PME Exchange		3	4	4		1	1

Table A.4—Continued

Questions/Specific Types of Data	Military Personnel Exchange Program				Operator Engagement Talks		
	Dept.	MAJCOM or NAF	Joint	Wing and Below	Dept.	MAJCOM or NAF	Joint
Request for Air Force Personnel Attendance		3	2	1		1	1
Request or Use of Air Force Aircraft		2	1	1			1
Summary of Statement of Intent		1	2	1			
Training Quotas		1	2				
Travel Orders		3	3	7			
Visit Request		4	3	5		2	1
Question 13: Do you collect data regarding attendees, participants, or numbers of graduates for the ?							
Question 14: If you answered "Yes" to the previous question, please describe the types of data you collect and the sources of the data:							
Country List	1		1			1	
Country Nomination	1	1	3	1		1	
Guest Lists			1	1		1	2
International Visit Request		5	1	2		1	2
Invitational Travel Orders		2	2				
Letter of Acceptance		1	2	2			1
Letter of Request			2	1		1	1
My own observations/trip report	1	3	1	6	1	2	2
Nomination Package	1	2	3	5		1	1
Proposal for PME Exchange		1	2	2			1
Request for Air Force Personnel Attendance		1	1			1	1
Training Quotas		1	1				
Travel Orders		2	2	4			
Visit Request		3	2	2		1	
Question 15: Do you collect data regarding participant views or observations, such as exit surveys for the ?							
Question 16: If you answered "Yes" to the previous question, please describe the types of data you collect and the sources of the data:							
After-Action Report		1	1	4	1	2	2
Annual Report		4	2	5		1	
End of Tour Report	1	7	2	13		2	
Meeting Minutes/Summary		2	1		1	3	3
My own observations/trip report		2	1	9	1	2	3
Progress Report			1	4		2	
Project Final Report			1		2	4	3
Project Quarterly Report			1	2			
Test and Disposition Report			1				

Table A.4—Continued

Questions/Specific Types of Data	Military Personnel Exchange Program				Operator Engagement Talks		
	Dept.	MAJCOM or NAF	Joint	Wing and Below	Dept.	MAJCOM or NAF	Joint
Training Report		2	1	6		1	1
Question 17: Do you collect data regarding capabilities assessments for the ?							
Question 18: If you answered "Yes" to the previous question, please describe the types of data you collect and the sources of the data:							
Exchange Agreement	1	1	1	5			1
Interim Tour Report	1	3	1	12			
International Agreement		1	1	1		1	
Memorandum of Agreement	1	1	2	3		2	1
Memorandum of Understanding	1	1	2	5		2	1
My own observations/trip report	1	1	1	16	1	3	1
Participant Entry or Exit Testing		1	1	1		1	1
Quid-Pro-Quo Analysis						1	1
Summary of Statement of Intent							
Question 19: Do you collect data regarding the effect of the ? on relevant partner capabilities?							
Question 20: If you answered "Yes" to the previous question, please describe the types of data you collect and the sources of the data:							
After-Action Report			1	3		2	1
Alumni Whereabouts		1	1				
Annual Report		1	1	2			
Certification to Congress							
End of Tour Report	1	2	2	15			
Interim Tour Report	1	2	1	11			
Meeting Minutes / Summary						1	2
My own observations/trip report	1		1	14		2	2
Progress Report			1	4		1	1
Project Quarterly Report	1	2	2	20			
Quid-Pro-Quo Analysis							1
Test and Disposition Report							
Training Report		1		2		1	
Question 21: Do you collect data regarding the funds expended for the ?							
Question 22: If you answered "Yes" to the previous question, please describe the types of data you collect and the sources of the data:							
Budget Allocation Memo		2	1	1		1	
Budget Projection	1	4	1	7		1	
Budget Request	1	3	1	7		1	
Loan Agreement							
My own observations/trip report	1			4			

Table A.4—Continued

Questions/Specific Types of Data	Military Personnel Exchange Program				Operator Engagement Talks		
	Dept.	MAJCOM or NAF	Joint	Wing and Below	Dept.	MAJCOM or NAF	Joint
Periodic Financial Report			2	3		1	
Quarterly Obligation Report	1			1		1	
Request for Fund Cite		3	1	3		1	
Travel Vouchers		3	1	11		2	
Question 23: Do you collect data regarding compliance with legal or regulatory requirements related to the ?							
Question 24: If you answered "Yes" to the previous question, please describe the types of data you collect and the sources of the data:							
Data Exchange Annex		2			1	2	1
Delegation of Disclosure Authority Letter	1	6		2	1		1
Extended Visit Authorization		6	1	2			
Information Exchange Agreement		3	1	3	1		1
My own observations/trip report	1	4	1	6	1		1
Request for Disclosure Authorization		6		1	1	2	1
Security Plan	1	6	1	2		1	
Question 26: Do you prepare reports that document specific activities or events (i.e., trip reports, after-action reports, surveys, etc.) for the ?							
Question 27: If you answered "Yes" to the previous question, please identify these documents:							
After-Action Report		1	1	9	2	3	
Annual Report		1	1	13		1	
End of Tour Report		3	1	31			
Interim Tour Report		1		24			
My own observations/trip report	1	4	1	14	2	3	3
Progress Report			1	2		1	
Project Final Report							
Quarterly Obligation Report				5		1	
Test and Disposition Report			1	2			
Training Report		1	1	8		1	
Question 29: Do you design, or contribute to the design of, specific events or activities for the ?							
Question 30: If you answered "Yes" to the previous question, please list the events or activities that you design, or contribute to the design of:							
Agenda	1		1	1	2	3	3
Background Material	1		1	6	1	4	1
Briefings	1		1	8	1	4	1
Budget Allocation Memo			1	1		1	
Budget Projection	1	1	1	4		1	
Budget Request		2	1	4		1	

Table A.4—Continued

Questions/Specific Types of Data	Military Personnel Exchange Program				Operator Engagement Talks		
	Dept.	MAJCOM or NAF	Joint	Wing and Below	Dept.	MAJCOM or NAF	Joint
Country List	1		1	2		1	
Country Nomination			2			1	
Exchange Agreement	1		2	5			1
Guest Lists				1	1	1	3
Master Data Agreement							
Master Information Agreement							
Position Description and Requisition Report	1			8			
Program Request						1	1
Project Proposal				1			1
Proposal for PME Exchange			1	5			1
Request for Disclosure Authorization		2		2		1	1
Request for Fund Cite			1	3			
Request for Use of Air Force Aircraft	1		1	1			1
Request for Air Force Personnel Attendance				1			1
Security Plan		3	1	3	1		
Summary Statement of Intent							
Test and Evaluation Plan			1	3			
Visit Request		3	1	4		1	2
Question 36: Do you collect data regarding cost of the overall program or the cost of a unit of output (i.e., one graduate, one event, etc.)?							
Question 37: If you answered "Yes" to the previous question, please identify this data:							
Budget Allocation Memo							
Budget Projection	1		1	5			
Budget Request	1	1	1	3			
Loan Agreement							
Periodic Financial Report			1	1			
Quarterly Obligation Report	1						
Request for Fund Cite				1			
Travel Orders		1	1	4			
Travel Vouchers		1		4			
Question 42: Do you gather data for the ? that reflects how well a specific event or activity met its objectives?							
Question 43: If you answered "Yes" to the previous question, please describe the type of data:							
After-Action Report				5		3	1
Annual Report		1		4			
Certification to Congress							

Table A.4—Continued

Questions/Specific Types of Data	Military Personnel Exchange Program				Operator Engagement Talks		
	Dept.	MAJCOM or NAF	Joint	Wing and Below	Dept.	MAJCOM or NAF	Joint
End of Tour Report	1	1	1	14			
Interim Tour Report	1	1	1	10			
Meeting Minutes/Summary						2	1
My own observations/trip report	1			10	1	3	2
Progress Report				2		1	
Project Final Report							
Project Quarterly Report							
Test and Disposition Report							
Training Report		1		5		1	1
Quarterly Obligation Report							

Conclusion

While not definitive, even as a representative sample from a larger collection of Air Force security cooperation programs, the results of the survey of MPEP and OET stakeholders do provide insights into potential stakeholder roles and data sources to support Air Force assessments. Using this approach will allow the Air Force to better understand its capacity to conduct assessments of its security cooperation programs, and will also highlight potential areas for improvement and enhancement.

Assessment Survey Template

This appendix provides a generic (i.e., non-program-focused) version of the RAND assessment survey in booklike format. The actual survey taken by Air Force security cooperation stakeholders during the summer of 2009 was in a web-based format that could be tailored to individual Air Force–managed programs, such as the MPEP.

Project AIR FORCE

Air Force Security Cooperation Request for Expert Feedback

Part I

Part I.A - Your Information

1. Please provide the information requested in the box below. This information will allow us to fully understand your organizations role as a stakeholder in the Air Force-managed security cooperation programs.

Privacy Act Statement for AF Surveys

AUTHORITY: 10 U.S.C.; 8013, SECAF

PURPOSE: The purpose of this survey is to obtain data from Air Force members about the ability of their offices/ units/ activities to collect information and support the assessment of Air Force Security Cooperation programs that are used to build partnerships, partner capability, or partner capacity.

ROUTINE USES: This information will be used as inputs to an analysis of the Air Force's preparedness to conduct assessments of its Security Cooperation programs. The results will help Air Force senior officials determine what additional steps, if any, the service should take in order to be able to perform the assessments required by the Guidance for the Employment of the Force (GEF).

DISCLOSURE: Participation is voluntary. No adverse action of any kind may be taken against any individual who elects not to participate in any portion of the survey. Personal identifying information is not used in any reports – only aggregate data will be reported.

1. Please enter your rank or grade:

2. Are you a contractor?
   ○ Yes ○ No

3. Please enter your organization and office symbol:

4. Please enter your position and title:

Part I

Part I.B – Air Force-managed Security Cooperation Programs

The Air Force manages a number of security cooperation programs that build partnerships, partner capability, and partner capacity. The programs we are currently studying are listed below.

a. Please review the list below carefully, as you may have a role in more than one program.
b. Please check each program in which you have a role. You will be asked to complete a separate survey for each program in which you have a role.

5. Please select each program in which you played a role. Each program is provided with a short description, for clarity.

☐ **Air and Trade Show:** Participation in international air shows and trade exhibitions allow the US to showcase its defense and weapon technology for the purpose of facilitating opportunities for cooperative research, development, and acquisitions.

☐ **Aviation Leadership Program:** Authorizes the participation of foreign and US military defense personnel in post-undergraduate flight training and tactical leadership programs in Southwest Asia without charge to participating countries.

☐ **Bilateral and Multilateral Forums:** Allows senior defense officials to participate in organizations that facilitate cooperation in military research, development, and acquisition between the US and its allies.

☐ **Bilateral/Regional Cooperation Programs:** Enables defense personnel of developing countries to attend bilateral or regional conferences, seminars or similar meetings that are in the national security interests of the United States.

☐ **International Cooperative Research and Development:** Allows for cooperative R&D projects on defense equipment and munitions with NATO and other friendly countries. The purpose is to improve common capabilities through emerging technologies.

☐ **Defense Research, Development Test and Evaluation Information Exchange Program:** Allows for reciprocal exchange of science and technology with allied and friendly nations. Explores future technology cooperation and multinational force capability.

☐ **Engineering and Scientific Exchange Program:** Allows for the reciprocal exchange of engineers and scientists between US and foreign countries to RDT&E facilities to increase cooperation and technical exchange in the R&D environment.

☐ **Foreign Comparative Test Program:** Authorizes the evaluation of defense equipment, munitions, and technologies developed by US allies and other friendly countries to determine their ability to satisfy US military requirements.

☐ **Latin American Cooperation (LATAM Coop):** Provides funds to conduct visits, exchanges, and seminars with Latin American countries to advance cooperation between the US and Latin American countries.

☐ **Military Academy Student Exchanges (US Air Force Academy):** Allows for the reciprocal exchange of cadets between the US Air Force Academy and foreign air force institutions.

☐ **Military Personnel Exchange Program (MPEP):** One-year exchange in equivalent grades and specialties with foreign nations. Enhances the ability of US military to perform coalition operations by developing international relationships.

☐ **NATO Forums:** NATO Forums advise the North Atlantic Council (NAC) on the development and procurement of equipment for NATO forces. Promotes standardization and cooperative research and information exchanges within the alliance.

☐ **Operator-to-Operator Talks Program:** Designed to enhance operator-to-operator relationships between the US Air Force and a select group of allies and partner air forces.

☐ **Professional Military Education Student Exchange:** Provides for no-cost, reciprocal professional military student exchanges.

☐ **UNIFIED ENGAGEMENT Regional DPC Seminars:** Designed to enhance relationships on a bilateral and multilateral level between the US Air Force and selected allies and partner air forces.

☐ **Competitions and Exercises:** Includes bilateral and multilateral exercises such as Red Flag, or competitions such as Rodeo, designed to enhance interoperability between the US Air force and selected allies and partner air forces.

6. Please indicate in this question which program that you selected in the previous question will be the subject of this survey response. If you selected more than one program, please return and complete a separate survey for each program. With your unique password, you can return to this online survey any time within our survey period to provide us information for each of the other programs for which you have a role.

○ Air and Trade Show

○ Aviation Leadership Program

○ Bilateral and Multilateral Forums

○ Bilateral/Regional Cooperation Programs

○ International Cooperative Research and Development

○ Defense Research, Development Test and Evaluation Information Exchange Program

○ Engineering & Scientific Exchange Program

○ Foreign Comparative Test Program

○ Latin American Cooperation (LATAM Coop)

○ Military Academy Student Exchanges (US Air Force Academy)

○ Military Personnel Exchange Program (MPEP)

○ NATO Forums

○ Operator-to-Operator Talks Program

○ Professional Military Education Student Exchange

○ UNIFIED ENGAGEMENT Regional DPC Seminars

○ Competitions and Exercises

Part II

Part II - Instructions

1. The questions that follow may be answered by clicking "Yes" or "No." If you select "Yes" for a question, you may be asked to provide additional information that elaborates on your answer.
2. **Stakeholder Roles:** The questions are grouped by four broad roles that a stakeholder might have with respect to a program: 1) Process Implementation, 2) Process Design and Development, 3) Making Recommendations, and 4) Making Decisions. You may have one or more roles in a program, so please complete each of the sections carefully. The roles are briefly described here:

    a. *Within each program, stakeholders have various levels of responsibility for carrying out activities. Some stakeholders **implement processes**, in other words, program managers assign them specific tasks which they then must carry out. These tasks might include organizing an event or providing subject matter expertise, establishing contracts or accounting for funds, or processing documentation required by Air Force instructions or other directives.*
    b. *Other stakeholders participate in the **design or development of processes**, carrying out such activities as developing lesson plans, contracts, or event agendas.*
    c. *Some stakeholders **make recommendations** to program managers about the size scope, or need for the program or a specific activity.*
    d. *Still other stakeholders **make decisions** regarding the specific activities, the need, or the scope of the program.*
3. Please answer each of the questions below as they relate to your duties with regard to the program for which you are providing the information. **In particular, please be sure to respond to follow-on »requests for specific information related to the roles, responsibilities and functions that you perform (e.g. events, activities, and types and sources of data).**
4. Please be as complete as possible, avoiding any acronyms or abbreviations.
5. After completing the questions for the first program you selected, you will be asked to repeat the questions for **each of the other programs** in which you have a role.

Part II.A - Process Implementation

7. Do you manage resources that are used in the implementation of the ?

    ○ No

    ○ Don't know/Not applicable

    ○ Yes

8. If you answered "Yes" to the previous question, please list the resources you manage:

    ☐ Facilities

    ☐ Funds

    ☐ Infrastructure

    ☐ People

    ☐ Other, please specify

Page 4

9. Do you directly observe or participate in any of the specific events or activities related to the ?

- ○ No
- ○ Don't know/Not applicable
- ○ Yes

10. If you answered "Yes" to the previous question, please list the events or activities in which you observe or participate:

- ☐ Command Post Exercise
- ☐ Competition
- ☐ Computer-Assisted Simulation or Gaming Activity
- ☐ Conference or Roundtable Discussion
- ☐ Education
- ☐ Field Exercise
- ☐ Personnel Exchange
- ☐ Table-Top Exercise
- ☐ Test of Field Experiment
- ☐ Training
- ☐ Other, please specify

Page 5

11. Do you collect data regarding requests to participate in the ?

○ No

○ Don't know/Not applicable

○ Yes

12. If you answered "Yes" to the previous question, please describe the types of data you collect and the sources of the data:

☐ Country List

☐ Country Nomination

☐ Guest Lists

☐ International Visit Request

☐ Invitation Travel Orders

☐ Letter of Request

☐ My own observations/trip report

☐ Nomination Package

☐ Program Request

☐ Project Agreement / Arrangement

☐ Project Nomination Form

☐ Project Proposal

☐ Proposal for PME Exchange

☐ Request for Air Force Personnel Attendance

☐ Request or Use of Air Force Aircraft

☐ Summary of Statement of Intent

☐ Training Quotas

☐ Travel Orders

☐ Visit Request

☐ Other, please specify

Page 6

13. Do you collect data regarding attendees, participants, or numbers of graduates for the ?

   ○ No

   ○ Don't know/Not applicable

   ○ Yes

14. If you answered "Yes" to the previous question, please describe the types of data you collect and the sources of the data

   ☐ Country List

   ☐ Country Nomination

   ☐ Guest Lists

   ☐ International Visit Request

   ☐ Invitational Travel Orders

   ☐ Letter of Acceptance

   ☐ Letter of Request

   ☐ My own observations/trip report

   ☐ Nomination Package

   ☐ Proposal for PME Exchange

   ☐ Request for Air Force Personnel Attendance

   ☐ Training Quotas

   ☐ Travel Orders

   ☐ Visit Request

   ☐ Other, please specify

Page 7

15. Do you collect data regarding participant views or observations, such as exit surveys for the ?

  ○ No

  ○ Don't know/Not applicable

  ○ Yes

16. If you answered "Yes" to the previous question, please describe the types of data you collect and the sources of the data:

  ☐ After-Action Report

  ☐ Annual Report

  ☐ End of Tour Report

  ☐ Meeting Minutes / Summary

  ☐ My own observations/trip report

  ☐ Progress Report

  ☐ Project Final Report

  ☐ Project Quarterly Report

  ☐ Test and Disposition Report

  ☐ Training Report

  ☐ Other, please specify

Page 8

17. Do you collect data regarding capabilities assessments for the ?

  ◦ No

  ◦ Don't know/Not applicable

  ◦ Yes

18. If you answered "Yes" to the previous question, please describe the types of data you collect and the sources of the data:

  ☐ Exchange Agreement

  ☐ Interim Tour Report

  ☐ International Agreement

  ☐ Memorandum of Agreement

  ☐ Memorandum of Understanding

  ☐ My own observations/trip report

  ☐ Participant Entry or Exit Testing

  ☐ Quid-Pro-Quo Analysis

  ☐ Summary of Statement of Intent

  ☐ Other, please specify

Page 9

19. Do you collect data regarding the effect of the on relevant partner capabilities?

    ○ No

    ○ Don't know/Not applicable

    ○ Yes

20. If you answered "Yes" to the previous question, please describe the types of data you collect and the sources of the data:

    ☐ After-Action Report

    ☐ Alumni Whereabouts

    ☐ Annual Report

    ☐ Certification to Congress

    ☐ End of Tour Report

    ☐ Interim Tour Report

    ☐ Meeting Minutes / Summary

    ☐ My own observations/trip report

    ☐ Progress Report

    ☐ Project Quarterly Report

    ☐ Quid-Pro-Quo Analysis

    ☐ Test and Disposition Report

    ☐ Training Report

    ☐ Other, please specify

Page 10

21. Do you collect data regarding the funds expended for the ?

    ○ No

    ○ Don't know/Not applicable

    ○ Yes

22. If you answered "Yes" to the previous question, please describe the types of data you collect and the sources of the data:

    ☐ Budget Allocation Memo

    ☐ Budget Projection

    ☐ Budget Request

    ☐ Loan Agreement

    ☐ My own observations/trip report

    ☐ Periodic Financial Report

    ☐ Quarterly Obligation Report

    ☐ Request for Fund Cite

    ☐ Travel Vouchers

    ☐ Other, please specify

Page 11

23. Do you collect data regarding compliance with legal or regulatory requirements related to the ?

    ○ No

    ○ Don't know/Not applicable

    ○ Yes

24. If you answered "Yes" to the previous question, please describe the types of data you collect and the sources of the data:

    ☐ Data Exchange Annex

    ☐ Delegation of Disclosure Authority Letter

    ☐ Extended Visit Authorization

    ☐ Information Exchange Agreement

    ☐ My own observations/trip report

    ☐ Request for Disclosure Authorization

    ☐ Security Plan

    ☐ Other, please specify

Page 12

25. Do you collect data regarding the implementation of the that was not mentioned above?

    ○ No

    ○ Don't know/Not applicable

    ○ Yes, please describe the types of data you collect and the sources of the data:

Page 13

26. Do you prepare reports that document specific activities or events (i.e., trip reports, after-action reports, surveys, etc) for the ?

    ○ No

    ○ Don't know/Not applicable

    ○ Yes

27. If you answered "Yes" to the previous question, please identify these documents:

    ☐ After-Action Report

    ☐ Annual Report

    ☐ End of Tour Report

    ☐ Interim Tour Report

    ☐ My own observations/trip report

    ☐ Progress Report

    ☐ Project Final Report

    ☐ Quarterly Obligation Report

    ☐ Test and Disposition Report

    ☐ Training Report

    ☐ Other, please specify

28. If you answered "Yes" to the previous question, please describe in greater detail, if necessary, and please indicate which offices and organizations review the indicated documents.

Part II.B - Process Design and Development

Part II.B - Process Design and Development

29. Do you design, or contribute to the design of, specific events or activities for the ?

   ○ No

   ○ Don't know/Not applicable

   ○ Yes

30. If you answered "Yes" to the previous question, please list the events or activities that you design, or contribute to the design of:

   ☐ Agenda

   ☐ Background Material

   ☐ Briefings

   ☐ Budget Allocation Memo

   ☐ Budget Projection

   ☐ Budget Request

   ☐ Country List

   ☐ Country Nomination

   ☐ Exchange Agreement

   ☐ Guest Lists

   ☐ Master Data Agreement

   ☐ Master Information Agreement

   ☐ Position Description and Requisition Report

   ☐ Program Request

   ☐ Project Proposal

   ☐ Proposal for PME Exchange

   ☐ Request for Disclosure Authorization

   ☐ Request for Fund Cite

   ☐ Request for Use of Air Force Aircraft

   ☐ Request for Air Force Personel Attendance

   ☐ Security Plan

   ☐ Summary Statement of Intent

   ☐ Test and Evaluation Plan

   ☐ Visit Request

   ☐ Other, please specify

Page 15

31. Do you develop documents such as Air Force instructions or other directives that guide or govern the conduct of activities or events within the ?

- ○ No
- ○ Don't know/Not applicable
- ○ Yes

32. If you answered "Yes" to the previous question, please list these documents:

- ☐ Data Exchange Agreement
- ☐ Exchange Agreement
- ☐ Information Exchange Agreement
- ☐ International Agreement
- ☐ International Program Directive
- ☐ Master Data Agreement
- ☐ Master Information Agreement
- ☐ Memorandum of Agreement
- ☐ Memorandum of Understanding
- ☐ Operating Procedures
- ☐ Position Description and Requisition Report
- ☐ Project Agreement / Arrangement
- ☐ Project Nomination Form
- ☐ Program Request
- ☐ Project Proposal
- ☐ Security Plan
- ☐ Test and Evaluation Plan
- ☐ Visitor Request
- ☐ Other, please specify

33. Is the design and development of activities or events for the governed by any Air Force Instructions, Manuals, or other directives?

   ◌ No

   ◌ Don't know/Not applicable

   ◌ Yes, please list these documents.

   [text box]

34. Is the design and development of activities or events for the governed by any non-Air Force Instructions, Manuals, or other directives (i.e., DoD Instructions, joint regulations, or any other non-Air Force directive)?

   ◌ No

   ◌ Don't know/Not applicable

   ◌ Yes, please list these documents:

   [text box]

35. Are there informal documents (e.g. continuity binders) that you refer to in the design and development of events or activities for the ?

   ◌ No

   ◌ Don't know/Not applicable

   ◌ Yes, please list these documents:

   [text box]

Page 17

36. Do you collect data regarding cost of the overall program or the cost of a unit of output (i.e., one graduate, one event, etc)?

　　○ No

　　○ Don't know/Not applicable

　　○ Yes

37. If you answered "Yes" to the previous question, please identify this data:

　　☐ Budget Allocation Memo

　　☐ Budget Projection

　　☐ Budget Request

　　☐ Loan Agreement

　　☐ Periodic Financial Report

　　☐ Quarterly Obligation Report

　　☐ Request for Fund Cite

　　☐ Travel Orders

　　☐ Travel Vouchers

　　☐ Other, please specify

　　[text box]

38. If you answered "Yes" to the previous question, please describe in greater detail, if necessary, and please indicate which offices and organizations review the documents identified above.

　　[text box]

Part II.C - Program Recommendations

> ## Part II.C - Program Recommendations

39. Do you make recommendations regarding the overall need for the ?

    ○ No

    ○ Don't know/Not applicable

    ○ Yes, please list the stakeholder(s) (i.e., office or organization) that receive(s) your recommendations:

    [text box]

40. Do you make recommendations regarding the need to increase or reduce participation in the ?

    ○ No

    ○ Don't know/Not applicable

    ○ Yes, please list the stakeholder(s) (i.e., office or organization) that receive(s) your recommendations:

    [text box]

41. Do you make recommendations regarding which countries participate in the ?

    ○ No

    ○ Don't know/Not applicable

    ○ Yes, please list the stakeholders (s) (i.e., office or organization) that receive(s) your recommendations:

    [text box]

Page 19

42. Do you gather data for the that reflects how well a specific event or activity met its objectives?

    ○ No

    ○ Don't know/Not applicable

    ○ Yes

43. If you answered "Yes" to the previous question, please describe the type of data:

    ☐ After-Action Report

    ☐ Annual Report

    ☐ Certification to Congress

    ☐ End of Tour Report

    ☐ Interim Tour Report

    ☐ Meeting Minutes / Summary

    ☐ My own observations / trip report

    ☐ Progress Report

    ☐ Project Final Report

    ☐ Project Quarterly Report

    ☐ Test and Disposition Report

    ☐ Training Report

    ☐ Quarterly Obligation Report

    ☐ Other, please specify

Page 20

44. Do you advocate for funds used to implement the ?

    ○ No

    ○ Don't know/Not applicable

    ○ Yes, please list the process(es) you use (i.e., PPBE, requests for O&M, ORF, or other existing funding sources, etc):

Part II.D - Program Decisions

Part II.D - Program Decisions

45. Do you determine the overall need for the ?
- ○ No
- ○ Don't know/Not applicable
- ○ Yes, please list process(es) you use, or participate in, to make this determination:

46. Do you set the objectives for the overall program?
- ○ No
- ○ Don't know/Not applicable
- ○ Yes, please indicate where the objectives are documented.

47. Do you set the objectives for specific events or activities within the ?
- ○ No
- ○ Don't know/Not applicable
- ○ Yes, please indicate where the objectives are documented.

48. Do you have access to information regarding other USAF programs security cooperation programs, such as their objectives, cost, and benefits?
- ○ No
- ○ Don't know/Not applicable
- ○ Yes, what are the sources for this information?

49. Do you have access to information regarding other USAF programs (not security cooperation) and the priority attached to each?
- ○ No
- ○ Don't know/Not applicable
- ○ Yes, what are your sources for this information?

50. Do you have access to information regarding other USAF programs (not security cooperation) and their objectives, cost, and benefits?
- ○ No
- ○ Don't know/Not applicable
- ○ Yes, what are your sources for this information?

Page 22

Part II E—Program Assessment Skills

Part II E—Program Assessment Skills

51. Do you believe that you, or personnel assigned to your position in the future, have/will have the skills to conduct appropriate security cooperation assessments (e.g. regarding the need for a program, program design, program compliance with policy, program outcomes, and program cost-effectiveness)?

    ○ No

    ○ Don't know/not applicable

    ○ Yes

52. If no, do you believe that you, or personnel assigned to your position in the future, would be prepared to conduct assessments if you had an appropriate checklist and set of instructions?

    ○ No

    ○ Don't know/not applicable

    ○ Yes

53. If no, do you believe that completion of a short course on conducting assessments (via classroom or online instruction) would be adequate preparation for someone in your position?

    ○ No

    ○ Don't know/not applicable

    ○ Yes

54. Please share with us any additional information you believe is pertinent to our inquiry or any comments you have about the instrument or our research.

Bibliography

AFI—*See* Air Force Instruction.

Air Force Instruction 16-107, *Military Personnel Exchange Program (MPEP)*, February 2, 2006.

Air Force Instruction 33-360, *Publications and Forms Management*, May 18, 2006.

Air Force Instruction 90-201, *Inspector General Activities,* November 22, 2004; Incorporating Through Change 3, July 19, 2007.

Air Force Policy Directive 90-1, *Policy Formulation*, August 1, 2003.

Berk, Richard A., and Peter H. Rossi, *Thinking About Program Evaluation*, Newbury Park, Calif.: SAGE Publications, 1990.

Defense Institute of Security Assistance Management, DISAM's Online Green Book, 2007. As of September 29, 2010:
http://www.disam.dsca.mil/pubs/DR/greenbook.htm

Defense Security Cooperation Agency (DSCA), Frequently Asked Questions (FAQs), 2007. As of September 29, 2010:
http://www.dsca.mil/PressReleases/faq.htm

Department of Defense Directive 5132.03, *DoD Policy and Responsibilities Relating to Security Cooperation*, October 24, 2008.

Department of Defense Directive 5230.20, *Visits and Assignments of Foreign Nationals,* June 22, 2005.

Dewitte, Lieven, "First New F-16 for Poland Peace Sky Program Accepted," General F-16 News, March 31, 2006. As of September 29, 2010:
http://www.f-16.net/news_article1728.html

DoD—*See* U.S. Department of Defense.

Hura, Myron, Gary McLeod, Eric Larson, James Schneider, Daniel Gonzales, Dan Norton, Jody Jacobs, Kevin O'Connell, William Little, Richard Mesic, and Lewis Jamison, *Interoperability: A Continuing Challenge in Coalition Air Operations*, Santa Monica, Calif.: RAND Corporation, MR-1235-AF, 2000. As of September 29, 2010:
http://www.rand.org/pubs/monograph_reports/MR1235/

Moroney, Jennifer D. P., Kim Cragin, Eric Gons, Beth Grill, John E. Peters, and Rachel M. Swanger, *International Cooperation with Partner Air Forces*, Santa Monica, Calif.: RAND Corporation, MG-790-AF, 2009. As of September 29, 2010:
http://www.rand.org/pubs/monographs/MG790/

Moroney, Jennifer D. P., Joe Hogler, Jefferson P. Marquis, Christopher Paul, John E. Peters, and Beth Grill, *Developing an Assessment Framework for U.S. Air Force Building Partnerships Programs*, Santa Monica, Calif.: RAND Corporation, MG-868-AF, 2010a. As of September 29, 2010:
http://www.rand.org/pubs/monographs/MG868/

Moroney, Jennifer D. P., Nora Bensahel, Dalia Dassa-Kaye, Heather Peterson, Aidan Kirby Winn, and Michael J. Neumann, *Enhancing the Effectiveness of the U.S. Air Force Operator-to-Operator Talks Program*, Santa Monica, Calif.: RAND Corporation, TR-805-AF, 2010b. Not releasable to the general public.

Mullen, Adm Michael G., remarks delivered at the Seventeenth International Seapower Symposium, Newport, Rhode Island, September 21, 2005.

Office of the Secretary of Defense, *Guidance for Employment of the Force*, Washington, D.C., July 18, 2008; not available to the general public.

Paul, Christopher, Harry J. Thie, Elaine Reardon, Deanna Weber Prine, and Laurence Smallman, *Implementing and Evaluating an Innovative Approach to Simulation Training Acquisitions*, Santa Monica, Calif.: RAND Corporation, MG-442-OSD, 2006. As of September 29, 2010:
http://www.rand.org/pubs/monographs/MG442/

Ratcliff, Ronald E., "Building Partners' Capacity: The Thousand-Ship Navy," *Naval War College Review*, Vol. 60, No. 4, Autumn 2007, p. 45.

U.S. Air Force, *Air Force Global Partnership Strategy: Building Partnerships for the 21st Century*, 2008.

U.S. Air Force Fact Sheet, "Air Force Inspection Agency," March 2009. As of September 29, 2010:
http://www.af.mil/information/factsheets/factsheet_print.asp?fsID=142&page=2

U.S. Department of Defense, *Security Assistance Management Manual (SAMM)*, DoD 5105.38-M, October 3, 2003. As of September 29, 2010:
http://www.dsca.mil/SAMM/

———, *Quadrennial Defense Review Report*, Washington, D.C., February 6, 2006a. As of September 29, 2010:
http://www.dod.mil/pubs/pdfs/QDR20060203.pdf

———, *Building Partnership Capacity: QDR Execution Roadmap*, Washington, D.C., May 22, 2006b. As of September 29, 2010:
http://www.ndu.edu/itea/storage/790/BPC%20Roadmap.pdf

———, *Guidance for Development of the Force: Fiscal Years 2010–2015*, April 2008; not available to the general public.

U.S. Department of State, United States Diplomatic Mission to Warsaw, Poland, ODC Activities 2009, 2009. As of September 29, 2010:
http://poland.usembassy.gov/poland/odc/odc-activities-2009.html

USAF—*See* U.S. Air Force.

Vick, Alan J., Adam Grissom, William Rosenau, Beth Grill, and Karl P. Mueller, *Air Power in the New Counterinsurgency Era: The Strategic Importance of USAF Advisory and Assistance Missions*, Santa Monica, Calif.: RAND Corporation, MG-509-AF, 2006. As of September 29, 2010:
http://www.rand.org/pubs/monographs/MG509/